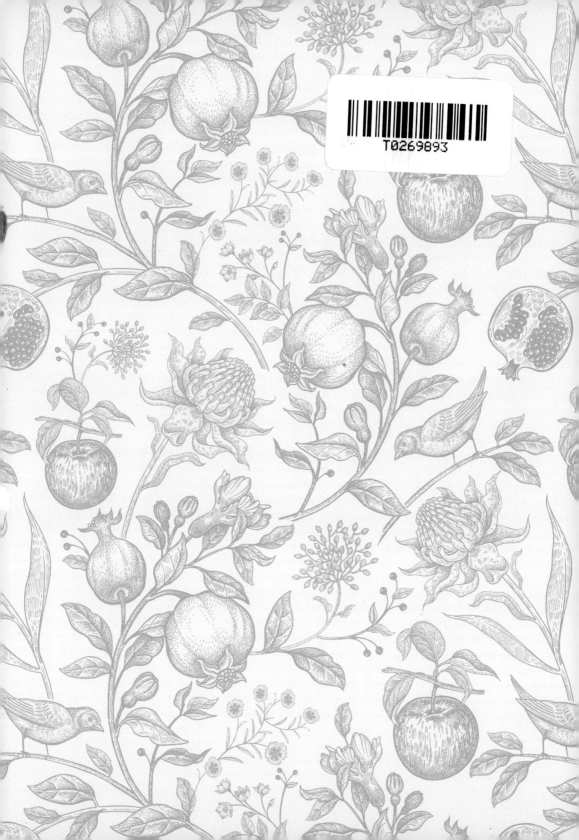

PRAISE FOR *Taming Fruit*

"A beautiful exploration of the life-giving bonds between trees, fruits, and people. Brunner is an astute guide to the fascinating reciprocal relationships between orchards and human culture."

DAVID GEORGE HASKELL, author of Pulitzer finalist *The Forest Unseen* and Burroughs Medal winner *The Songs of Trees*

"An enchanting journey through the world of orchards and botanical curiosities. Beautifully illustrated and written with infectious and cultured enthusiasm, anyone who is even a tentative gardener will cherish this lovely book."

BRIAN FAGAN, author of *The Little Ice Age* and *The Intimate Bond*

"This rich combination of glorious illustrations with cultural history, botany, anthropology, and personal anecdote will enthrall and delight anyone curious about the origins of orchards and the fruit they bear."

HELENA ATTLEE, author of *The Land Where Lemons Grow*

"Fruit was there at the beginning of the human story, Bernd Brunner argues in this crisply written and lushly illustrated book, and it's been with us ever since—in birth and death, peace and war, art and myth, science and religion."

ZACH ST. GEORGE, author of *The Journeys of Trees*

"From American cider orchards to Mediterranean citrus groves, this beautifully illustrated book is an enticing insight into the world of fruit trees. Brunner's eloquent and engaging account reminds us that the magic of the orchard extends far beyond its fruit."

LEIF BERSWEDEN, author of *The Orchid Hunter*

"A beautifully illustrated journey through different lands and times, Bernd Brunner's *Taming Fruit* enlightens us on the deep and winding history of how humans have used fruits and capitalized upon their sweetness and delight for our palates!"

NEZKA PFEIFER, museum curator, Stephen and Peter Sachs Museum, Missouri Botanical Garden

"Bernd Brunner's fantastic book opens our eyes for the orchard as a way of life in which nature and culture coexist. I'm now dreaming of the world as one gigantic orchard, teeming with life."

CHRISTIAN SCHWÄGERL, award-winning environmental journalist and author of *The Anthropocene*

"*Taming Fruit*'s fascinating tales, paired with gorgeous historical art, are potent lessons in cultivation that we can imitate today—for sustainability, freshness, and the joy of eating one's own peach or olive."

ERICA GIES, environmental journalist

Also by Bernd Brunner

Winterlust: Finding Beauty in the Fiercest Season
Birdmania: A Remarkable Passion for Birds

TAMING FRUIT

How Orchards Have
Transformed the Land,
Offered Sanctuary,
and Inspired Creativity

BERND BRUNNER

Translation by LORI LANTZ

GREYSTONE BOOKS
Vancouver/Berkeley

Greystone Books Ltd.
greystonebooks.com

Cataloguing data available from Library and Archives Canada
ISBN 978-1-77164-407-5 (cloth)
ISBN 978-1-77164-408-2 (epub)

Editing by Jane Billinghurst
Copy editing by Jess Shulman
Proofreading by Jennifer Stewart
Cover design by Belle Wuthrich and Jessica Sullivan
Text design by Belle Wuthrich
Front cover images: bird from mamita / Shutterstock;
trees and fruit from Adolphe Philippe Millot, *Nouveau Larousse Illustré*
(Paris: Librairie Larousse, 1933); peach from Charles Mason Hovey,
The Fruits of America (New York: D. Appleton, 1853)

Printed and bound in China by 1010 Printing International Ltd.

Greystone Books gratefully acknowledges the Musqueam, Squamish,
and Tsleil-Waututh peoples on whose land our office is located.

Greystone Books thanks the Canada Council for the Arts, the British Columbia
Arts Council, the Province of British Columbia through the Book Publishing Tax
Credit, and the Government of Canada for our publishing activities.

— CONTENTS —

— PROLOGUE —

The Seeds for This Book

ACCORDING TO HENRY DAVID THOREAU, "When man migrates he carries with him not only his birds, quadrupeds, insects, vegetables, and his very sword, but his orchard also." Efforts to cultivate fruit trees have historically connected regions and continents—and continue to do so today. As a result, they involve the interplay of time periods, landscapes, and nations. This book provides an overview of the different types of orchards that have existed throughout history and the principles by which they were organized. After all, the form that an orchard takes reflects the conditions of the time in which it was created. I will also endeavor to paint a picture of the life and work that took place among the trees, along with the thoughts that they inspired.

Places where various plants (and trees) are grown are often divided into two categories: ornamental spaces that fulfill an aesthetic purpose and productive spaces where the emphasis is on the harvest. In this way of seeing, ornamental gardens are works of art, while those where luminous fruit swells beneath the canopy of leaves are the products of labor. Is this really the case? Can't an orchard—at least one not cultivated on an industrial scale—be beautiful? In this book, we will discover gardens and orchards that blur the lines between these supposedly clear categories. After all, many ways exist to shape these spaces: the interplay of light and shadow, paths that unfurl before the wanderer, places to sit, perhaps a little hut offering shelter from sudden rain, a swing.

But no matter how well designed, how much care is lavished on it, or how productive it might be, an orchard is by nature impermanent, even though it may exist near a human settlement for decades or more. As soon as fashions change, food is sourced from elsewhere, or the owners move away and no one feels responsible, other plants begin to take over. Eventually, all signs of the orchard disappear. But even when lost orchards no longer appear on any map, they did exist. They have a history.

Perhaps it makes sense to think about an orchard as a kind of stage—one where a highly specific drama plays out between fruit trees and their caretakers, whoever they may be. Viewed in this way, orchards invite us to enjoy the complex spectacle of fruit growing and ripening in the company of animals, people, and other plants.

THE IMPETUS FOR THIS book came from an article that I discovered several years ago in a recent French book about the history of fruit cultivation. One of the many topics covered was Gesher Benot Ya'aqov, an archaeological site in the northern Jordan valley. Researchers there discovered stone tools and a variety of organic remains, among them various fruits and nuts, including acorns,

previous page
When the Dutchman Pieter Hermansz Verelst painted this young lady in the seventeenth century, it was fashionable to pose with fruit.

left
The avocado or
"alligator pear," early
nineteenth century.

almonds, water chestnuts, and Mount Atlas mastic (*Pistacia atlantica*), an evergreen shrub related to the pistachio.

The remains found at Gesher Benot Ya'aqov were estimated to be about three hundred thousand years old—a number so seemingly unbelievable that I had to read it several times. It meant that the finds came from the Paleolithic era, roughly one hundred thousand years before *Homo sapiens* likely arrived from the African savannas. At that time, half of Europe and North America were buried under a layer of permafrost. What is more, as more recent research suggests, some of these remains might be considerably older than that.

A look at a map confirmed my vague suspicion: as luck would have it, I had firsthand knowledge of the area. At one point in the mid-1980s I found myself near the Sea of Galilee in northern Israel. I spent a few weeks at the kibbutz Ami'ad (which means "my people, forever") above the sea, not far from Jordan and the Golan Heights. The Gesher Benot Ya'aqov site is just over six miles (ten kilometers) away.

The fruit trees grown at Ami'ad were not native to the region. I was assigned to help with the harvest of one such import, the avocado. This highly nutritious pear-shaped fruit originated in the forests of Mexico and spread from there to Brazil, probably with the help of now-extinct giant ground sloths. Archaeological evidence shows that people were already using them around six thousand years BCE, but first began actively cultivating them about a thousand years later. Because of the fruit's reptile-like skin, English speakers originally called avocados "alligator pears."

The kibbutz's avocado grove of at least a few hundred trees was located outside the main settlement. The six-foot (two-meter) trees with their wide-reaching branches stood lined up in rows, as far apart as the trees were tall. Most of the fruit could only be reached with a picker, and here and there we had to climb up into the branches. We fought our way through the thick, dark-green foliage of the crooked, twisted trees to pull—or often, twist—the still rock-hard avocados from the branches. The green buttery flesh of their ripe counterparts was served at dinner every night. I quickly grew tired of eating avocados and the many calories took their toll, but there simply weren't many other options. This overabundance is a common theme through orchards in history and, as I was to discover, people came up with ingenious ways of processing fruits and nuts to vary their taste and texture, and to preserve them as year-round sustenance—but despite these efforts often also found themselves with diets as monotonous as mine.

The tantalizing information in the French article mentioned earlier seemed like a sign to investigate further. I was fascinated by how very long ago someone had gathered the ancient fruits and nuts found at this site. Although we can't be certain which group of early humans left these remains (*Homo erectus*, *Homo heidelbergensis*, or even, perhaps, soon-to-be Neanderthals), what we do know is that even back then in the Lower Paleolithic, our ancient ancestors were picking and processing bounty from the wild.

When I contacted the author of the article, archaeobotanist George Willcox, about this site, he pointed out that the Mount Atlas mastic enjoyed by our distant ancestors continues to be eaten and used today by people in Syria, Turkey, and beyond. After the Stone Age, it was one of those fruit-bearing trees that provided much more to the communities that valued it than simply food. It played and continues to play its part in both commerce and pleasure. Its sap can be processed into alcohol, medicines, perfumes, and incense. Its bark contains tannins for processing animal hides, and its sturdy root system continues to be important in preventing soil erosion in dry and dusty parts of the world. The fruit of another species in the genus—*Pistacia vera*—are the pistachio nuts that we are familiar with. In eastern Turkey, *Pistacia vera* trees are grafted on *Pistacia atlantica* rootstock, which is local and sturdier.

I BEGAN TRACING THE history of orchards because I wanted to better understand the coevolution of fruit trees and humans. This shared process changed both participants. Obviously, eating delicious fruit enhanced humans' meals and thus their lives. As a result, people influenced the structure of the trees and their ability to produce desirable fruit, making them even more attractive. And beyond the trees and fruit themselves, humans are connected to the land where their orchards grow—land where they not only planted, watered, and harvested but also conversed, lived their lives, and enjoyed themselves.

Based on everything we know about the origins of agriculture, the cultivation of fruit trees often went hand in hand with a settled home in the vicinity. Marked off and secured as a productive piece of land, it became part of the property of a specific family or clan. And wherever the precious trees and bushes grew, their owners developed practices for gathering the bounty they produced. They plucked fruit from the trees; combed berries from the bushes; shook apples, cherries, or plums from the branches; or even beat the trees to get them to release nuts and olives into nets or onto the ground. With a little imagination, the rustle of wind in the leaves as you walk under fruit trees or through olive groves even today evokes the bustle of these early communities as they feasted on this fresh harvest or processed it into oil or dried fruit that would nourish them through more barren seasons.

THROUGH THE AGES, THE biological growth of fruit has been accompanied by a long historical development that can rightfully be compared with the domestication of dogs, cattle, or chickens. Michael Pollan formulated an intriguing theory along these lines: not only has human cultivation changed plants, but plants have affected us as well—in a process that almost seems conscious.

The Egyptian botanist Ahmad Hegazy and his British colleague Jon Lovett-Doust pursue this idea further, asserting that *as far as plants are concerned, we are simply one of thousands of animal species that more or less unconsciously "domesticate" plants. In this co-evolutionary dance (and in common with all animal species including humans), plants must spread their offspring to areas where they can thrive and so pass on their genes.*

They go on to argue that

in the game of evolution for crop and garden-ornamental species, humans have selected and bred plants for their desired attributes—including size, sweetness, color, scent, fleshiness, oiliness, fiber content, and drug concentration.

In *On the Origin of Species*, Charles Darwin described a mechanism applied "almost unconsciously" by generations of past orchardists and gardeners:

It has consisted in always cultivating the best known variety, sowing its seeds, and, when a slightly better variety chanced to appear, selecting it, and so onwards.

The results are works of art created through the combined effort of countless people, toiling in forgotten orchards in the depths of history, always working in an alliance with the forces of nature. Seen in this light, fruit is a generous offering provided by trees—to all animals, who have also contributed to the selection of especially useful or enjoyable varieties, increasing their value—and of course to humankind.

— 1 —

Before
There Were
Orchards

MILLIONS OF YEARS AGO, when the continents were settling into the configuration we know today and ice covered large portions of the Northern Hemisphere, many regions that are now temperate still had the characteristics of a tundra. In those times, diminutive wild cranberries, strawberries, raspberries, and blueberries grew across the landscape. Relatives of the fruit and nut trees that would later grow in the temperate regions of the North—apples, pears, quinces, plums, cherries, and almonds—were also abundant. During the short summers, these wild fruits were a sought-after source of additional nutrition for all types of animals, from the smallest insects to birds and reptiles and on to the largest mammals.

The fruit itself, especially the aromatic, fragrant, often juicy, and more-or-less sweet layer surrounding the seeds, was originally nothing more than a trick for attracting animals so that they would bring the seeds to another location where they could grow and the plant, as a result, would spread. Before early humans appeared, animals helped power a process of natural selection among fruit-producing plants. For example, birds prefer sweet berries to sour ones, so over time it was the seeds of the ripe berries—in other words, those capable of sprouting—that spread the most.

Several years ago anthropologists from New York University formulated an interesting thesis that fruit played a much larger role in evolution than previously believed. Alexandra DeCasien, a primatologist involved in the study, claimed that primates with diets consisting at least partially of fruit have significantly larger brains than those that eat leaves only. The scientists' findings suggest that this is because fruit-eating animals must search more intensively for their food and know their way around the forest—in other words, they depend more on their cognitive abilities. In fact, animals that eat fruit instead of leaves have brains that are 25 percent heavier relative to their body weight. Earlier, a group of scientists working with anthropologist Nathaniel Dominy from Dartmouth College in Hanover, New Hampshire, discovered that chimpanzees can apparently determine whether fruit is ready to eat just by pressing it with their fingertips.

previous page
Sour apples from a late fourteenth-century Italian edition of *Tacuinum Sanitatis (Maintenance of Health)*.

below
Monkeys are known for their appetite for fruit, 1857.

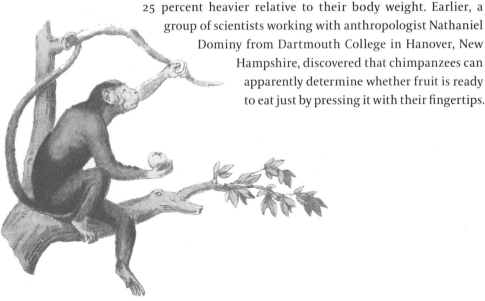

The act of searching for ripe fruit on branches, along with needing to know where to find trees that produce fruit, what time of year fruit ripens, and how to free the fruit from its sometimes hard shell—all activities requiring heightened mental power—may have contributed to bigger brains. Demands on the physical and cognitive abilities of monkeys and apes are much higher than on those of animals that only eat grass, for example. Before this connection was understood, scientists believed that social interaction was the main driver of brain development.

above
Cherry tree from the late fourteenth-century Italian edition of *Tacuinum Sanitatis (Maintenance of Health)*, originally published in Arabic in the eleventh century.

BOTANISTS EXPEND MUCH TIME and thought on defining what constitutes a fruit. I propose taking the perspective of the fruit's end user, that is to say the one who enjoys it. The word *fruit*, then, applies to the parts of plants that grow on trees, shrubs, or small bushes and that, over the course of history, have been incorporated into the human diet. Some types of fruit are fleshy with a pit or stone at the center, such as the cherry, plum, and olive. Then there are fruit species with seeds or pips rather than stones—apples, pears, and grapes—and small, soft fruit such as strawberries and black currants. A book on orchards would not be complete without also mentioning nuts and a most remarkable fruit that is formed from a cluster of inverted flowers: the fig.

Fruit is an important element in our daily diet. It contains a variety of vitamins and minerals, enzymes, and other substances that are essential for our health and well-being. One of these is vitamin C. Like tarsiers, monkeys, and apes, humans are members of the suborder of dry-nosed primates or haplorrhines. These primates and several other mammals—specifically, bats, capybaras, and guinea pigs—form an unusual club: they all need to consume vitamin C (also known as ascorbic acid) because their bodies can't produce it on their own. Although ascorbic acid was only identified about a hundred years ago, people have long recognized that eating fruit regularly is good for you. The traditional expression "An apple a day keeps the doctor away" is evidence of this intuitive knowledge—although today scientists know that two apples a day are better than one.

The high nutritional value of fruit is not the only reason it attracts us. We are drawn by its beautiful colors and interesting shapes, and the experience of eating fruit is also rewardingly complex. Its aroma, its sweet or sour taste, the texture of its flesh, the amount of water it contains and the resulting dry or juicy sensation: it all adds up to an impression that draws people back again

left
A selection of citrus (from left to right): citron, small Chinese orange or wampee (below the citron), sweet orange, and lemon, 1868.

and again. Our ancestors learned early on which parts of plants taste good and which are inedible or even poisonous. And plants, fruit, roots, and seeds have the advantage of being user-friendly. Berries, for example, can be easily gathered by hand and don't need to be prepared or processed in any way before being eaten. Fruit can usually be consumed raw. And coloring provides a clue to whether fruit is ready to be enjoyed—an indicator that you and I notice better than most mammals, who can't tell the difference between red and green.

Although most types of fruit were originally quite small, they were worth seeking out simply because they could be gathered without the risks of physical injury that hunting entails and added some variety to a limited spectrum of nutritional options. People naturally frequented the trees with the fruit that tasted best, and at some point had the idea to take some of the seeds and plant them closer to their homes. Eventually early orchards—perhaps consisting of just a few trees—were planted in the vicinity of human settlements, and later within the land claimed by particular families or clans. The initial transition to cultivation probably took place in areas like river valleys and oases, where the trees had already gained a foothold naturally.

At a certain point—one that varied from region to region—people understood that by selecting and breeding particular trees they could influence the resulting fruit. We should remember that growing fruit and collecting it in the wild have never been mutually exclusive activities. The ongoing selection of new types that emerged from wild fruit trees growing in the forest or transplanted to the outskirts of human habitations gradually produced the predecessors of orchards, although these early varieties were still primitive and had little in common with the fruit we know today.

___
left
Ripe figs with characteristically shaped leaves, eighteenth century.

Most of the non-tropical fruit we cultivate originated in places with a wide variety of wild fruit species that, in turn, display a varied genetic makeup. Nikolai Vavilov, a Russian botanist (1887–1943), put forward the hypothesis that the "homeland" of a plant species is the place where it exhibits the greatest variation. This genetic diversity meant that countless instances of cross-breeding were possible within a single species of wild fruit—and that this cross-breeding took place as part of an entirely natural process, with no need for any human interference.

As the genes were remixed, plants and fruit resulted with slight differences from their forebears. Larger hybrids were more likely to be sought out and have their seeds distributed by animals. Centers of such genetic activity are largely located in regions with a Mediterranean or subtropical climate, especially near the Mediterranean itself, the Middle East, Southwest to central Asia, the Indian subcontinent, and East Asia. Large parts of Africa, South America, and Australia are also genetic hotspots.

IN 2006, A STUDY published by a team of American and Israeli scientists made considerable waves in the field of archaeobotany. In the lower Jordan valley they found six small figs that appeared to have been cultivated—in other words, intentionally planted. The researchers determined that these remains were 11,200 to 11,400 years old. This discovery turns the usual sequence of agricultural development—first grain, then fruit—on its head, at least in this documented case. But this tantalizing find cannot tell us how this group of trees was organized or what the resulting space would have looked like.

At this point nothing would please me more than to present you with a historically documented account of the world's *very first* orchard: the lovely trees and delicious fruit, along with the people and animals who spent time there. Sadly, I can't. But a few facts about the development of orchards are fairly certain and can give us an idea.

Was it the fig, the olive, the date, or the pomegranate that first inspired people to plant their own group of fruit trees? The biggest problem in answering this question is that while carbonized remains of all these species exist, it is generally not possible to tell whether the specimens in question represent wild or cultivated varieties. This obstacle remains even when scientists can determine the age of a find fairly precisely. The transformation from a wild to a cultivated plant takes an enormously long time and its physical signs emerge extremely slowly.

Yet one reliable marker of cultivation does exist: remains discovered well outside the area where a plant was typically found indicate that these specimens were intentionally planted and given water that would not have reached them naturally. One example is provided by the olive pits and remnants of olive wood found just north of the Dead Sea.

The earliest evidence that humans harvested olives from wild trees comes from a similar geographic area and dates back to the Epipaleolithic, the transitional period between the Early and Late Stone Age (15,000 to 10,000 BCE). For many years, the broad consensus among archaeobotanists was that humans first cultivated olive trees about six thousand years ago in an area north of the Dead Sea and south of the Sea of Galilee, in the Jordan valley. However, more recent studies using improved analytical methods show that early olive cultivation was happening in many parts of the Mediterranean region, the Middle East (including Cyprus), the Aegean region, and around the Strait of Gibraltar. These ancestral wild gene pools provided the essential foundation for olive cultivation and breeding over a wide area.

Is it possible to more narrowly define where the cradle of olive domestication was located? In the central Euphrates valley, near the current border of Turkey and Syria, researchers have found remnants of olive wood and seeds from trees cultivated during the Bronze Age. One great advantage of olive trees is their ability to grow in even relatively low-quality soil. And the Bible's many references to both olive trees and olive oil illustrate the fruit's importance throughout the Middle East. For example, Psalm 128:3 contains a memorable comparison:

Your wife will be like a fruitful vine
Within your house,
Your children will be like olive shoots
Around your table.

Extensive olive groves continue to exist in the Middle East. Abandoned oil mills—sometimes just fragmentary ruins—are still being found in fields that have long gone back to nature. The oil they produced was used as an ointment, fuel for lamps, and a solution for fragrances in perfumes and cosmetics.

Olive groves can also be found in the very far west of Europe, in Alentejo in Portugal: endless olive groves stretching over the ochre-colored earth, often rising and falling along rolling hills. The only interruptions are scattered farmsteads with whitewashed buildings gleaming in the sun. There is little to hear besides twittering birds, the measured sounds of a cicada's song, or the beat of a mule's hooves. The characteristically sweet, heavy scent of alpechin, the wastewater that results from the olive-milling process, lies over the land.

Olive trees often reach an astonishingly old age. For example, the gnarled *oliveira do Mouchão*, the Mouchão olive tree, is estimated to be 3,350 years old. It stands in Abrantes in central Portugal, just over half a mile (one kilometer) as the crow flies from the Tejo river. At its base, its circumference measures 36.75 feet (11.2 meters). Olive trees of a similarly venerable age have also been identified on Sardinia, in Montenegro, and in Greece. At this point, Mort Rosenblum, the "biographer" of the olive, should have the last word on this topic: "Olives were domesticated before anyone devised words to record the fact."

— 2 —

The Rustle
of Palm
Leaves

SHADY PALM GROVES HAVE provided welcome spots for nomads to rest since time immemorial. Date palms in particular played an important role for people in ancient Egypt, the wide steppes of Syria, the deserts of the Arabian Peninsula, and Mesopotamia. Graffiti showing palms, people, and animals was discovered on the cliffs of Jabal Yatib, in the Ha'il region of Saudi Arabia, and apparently dates from the Bronze Age. And Neolithic clay pots decorated with palm fronds indicate that humans used these trees in that earlier period as well. The earliest mummies, found in the Nile valley, were wrapped in mats made from palm leaves.

For *Homo sapiens* in places like the Olduvai Gorge in today's Tanzania, oases must have been essential to surviving dry periods. Nearly one hundred years ago, the Australian-British archaeologist Vere Gordon Childe posited the "oasis theory." He suggested that humanity's transition to an agricultural way of life took place in oases and river valleys because at the end of the Pleistocene—a period of extreme drought, in which plants and animals were scarce—people sought refuge in these locations. While the notion that agriculture was born in the oases has since been refuted, they have certainly long played a central role for humankind.

A number of different factors would have determined whether people settled in them—in other words, whether continuously inhabited structures existed there or the time spent at these spots represented just a temporary station in a primarily nomadic life. In any case, the palm—in the role of a plant in whose shadow other trees or bushes might grow—likely acted as a cornerstone of such plant collectives. These groves probably extend back in time before the early civilizations that developed around the Fertile Crescent, a (not surprisingly) crescent-shaped region of the Middle East where some of the world's earliest traces of advanced cultures have been found.

Oases were often located on ancient trading routes and helped make movement over long distances possible—we can imagine them as staging posts where travelers could refill their supplies of fresh water and fruit. They were strung out along the Darb el-Arbain camel route stretching from central Egypt to Sudan, the Incense Route from southern Arabia (today's Oman and Yemen) to the Mediterranean, the route from Niger to Tangier in northern Morocco, and not least, critical stretches of the legendary Silk Road that linked one side of the Eurasian continent with the other.

previous page
Taking a nap under a palm tree in the Canary Islands, 1960.

___

left
The oasis: where palm trees, water, camels, and Bedouins meet, early twentieth century.

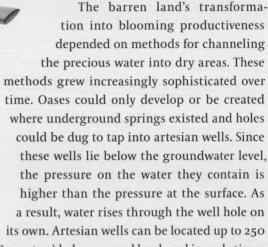

The barren land's transformation into blooming productiveness
depended on methods for channeling
the precious water into dry areas. These
methods grew increasingly sophisticated over
time. Oases could only develop or be created
where underground springs existed and holes
could be dug to tap into artesian wells. Since
these wells lie below the groundwater level,
the pressure on the water they contain is
higher than the pressure at the surface. As
a result, water rises through the well hole on
its own. Artesian wells can be located up to 250
feet (80 meters) below ground level, and in early times
they could only be accessed with a pickax. A basket of woven palm
leaves and a rope made of palm fiber helped to bring up the loosened sand or dirt. All in all, the technical challenges in cultivating
oases are considerable. Furthermore, the lack of water in general is
complicated by the tendency of the little water available to evapo-
rate quickly, leaving salt that wicks further moisture from the
plants and harms them.

An old saying asserts that date palms grow best "when they
stand with their feet in the water and their heads in the burning
sun." The roots of full-grown trees can extend deep enough to
reach the groundwater. It may seem paradoxical, but palm roots
actually resemble those of swamp-dwelling or aquatic rushes.
Furthermore, despite the very different conditions in which these
plants grow, they both have fibrous matting root systems that prevent soil erosion—one in a very dry environment and the other in
a very wet one. In terms of their structure, palms, like rushes, are
monocots; in other words, they're more akin to grasses than to

trees. Palms are unique among monocots in their ability to thicken their stems and grow tall. They are surprisingly undemanding when it comes to water quality, undisturbed even by very salty or brackish water. Cool temperatures are their undoing, however: as soon as the thermometer drops below 44 degrees Fahrenheit (7 degrees Celsius), they stop growing.

The great naturalist Alexander von Humboldt once referred to palms as "the loftiest and most stately of all vegetable forms." While differences clearly exist among the more than two hundred genera and nearly three thousand species of palms, all members of this family share certain characteristics: a crown of fronds with narrow leaflets arranged on both sides of a central stem—sometimes larger and sometimes smaller—emerging from the tip of a slim, branchless trunk.

According to legend, the date palm was born somewhere between the Tigris and the Euphrates from a mix of heavenly fire and earthly water. Research has since thrown not only this questionable origin into doubt, but its location as well. Today scientists believe that the world's first palms stood on the eastern coast of the Arabian Peninsula. This theory is based on cuneiform inscriptions found in the ruins of the Sumerian capital of Ur that describe how the trees had already been cultivated long before. These accounts relate that the people of the mythical paradise of Dilmun brought palm trees with them when they left their land for Mesopotamia. (Dilmun, supposedly the seat of an advanced culture, is described as located between today's countries of Kuwait and Qatar and was presumably therefore the island of Bahrain.)

THE BOTANICAL ORIGIN OF the cultivated date palm has not been determined with certainty. The tree could be a descendant of a wild palm that no longer exists, but it could also be related

to a group of palms that still grow today in a variety of far-flung locations. This group includes the rare *Phoenix atlantica* on the Cape Verde islands; the slender *Phoenix reclinata* or Senegalese date palm found in tropical Africa, on Madagascar, and in Yemen; and the *Phoenix sylvestris* at home in Pakistan, India, and the rest of Asia. All fourteen species of the genus *Phoenix* can reproduce with one another, and the only main difference between them is the location where they grow.

In North Africa and Arabia, the fruit of *Phoenix dactylifera*—packed with easy-to-digest sugars, proteins, minerals, and vitamins—has traditionally been an important source of nutrition for humans and animals alike. Countless dishes and desserts incorporate fresh or cooked dates. The fruit is often processed on site at the oases or plantations where it grows and cooked in large copper kettles with water to a puree that is used to sweeten all kinds of foods.

Commentators as early as the Greek geographer Strabo knew that parts of palm trees could be used to make wine, vinegar, honey, flour, and various fibers and mats. The date pits were burned as fuel or fed to animals. And in southern Mesopotamia, where wood was scarce, the trunks were an extremely desirable material, used to build farming implements, furniture, and ships for war or trade. Traditionally, farmers who worked in the oases used palm fronds to build huts. They were ideal for this purpose because they were long—over twelve feet (four meters)—and could be woven to form a structure that provided shade but still allowed the air to circulate.

People apparently understood early on how to best encourage palms to reproduce. These plants can stretch nearly one hundred feet (thirty meters) into the heavens. Pollination takes place thanks to the wind or, less frequently, insects carrying pollen from the male blossoms to the female ones. As the two types of blossoms grow on separate trees, relying on air currents to bring about their union is a risky business. To ensure that as many blossoms were pollinated as possible, specialists spent the period from January to March painstakingly rubbing the flowering stems of the female blossoms with male blossoms plucked from the trees. The more thoroughly they performed this laborious task—which the Greek scholar Theophrastus once described as "dusting"—the larger the subsequent date harvest. And that harvest can be quite large indeed: a single tree can produce up to 300 pounds (140 kilograms) of dates. With that in mind, it's easy to understand why dates were known as "the bread of the desert." Assyrian bas reliefs in Mesopotamia often depict scenes of gardeners dusting the trees to pollinate them. They had an incentive to do a thorough job as, in the Assyrian period, gardeners or plant dusters shared in the orchard's success: they received a third of the harvest.

An interesting dilemma for those harvesting dates was the fact that when the fall harvest came not all dates in the same bunch had

left
An anonymous poet once declared the date palm to be the queen of trees and dates to be the fruit of heaven, 1821.

reached the same state of ripeness. In the natural course of things, different rates of ripening meant that fresh dates were available for eating over a longer period, but this advantage was an obstacle to efficient commercial use. However, people discovered a way to limit losses at harvest time: picking the fruit well in advance and letting it ripen slowly in the shade. They also learned that making small cuts in the skin encouraged the ripening process.

Another way of tricking nature and increasing the fruit harvest was to only plant female trees, since they are the ones that bear fruit. If the trees reproduce naturally—as they do in the uncultivated oases—half of the palms that develop from fallen fruit or pits are male. Furthermore, young plants that emerge from date pits require a great deal of water if they are to survive to grow tall. At a certain point, oasis dwellers realized that they could avoid this issue by taking cuttings from the trunk of a grown female tree and planting them, which then grow into new female trees. This procedure shortens the period in which the young trees bear no fruit to four to five years, and more fully grown date palms can withstand longer dry periods. The male plants, in turn, are cultivated at a different location for their pollen, an important crop in its own right that finds buyers at markets throughout the Arabian region. In addition to being critical for fruit cultivation, date pollen has long been believed to aid fertility and help cure a variety of illnesses.

As the tallest-growing plants in the oases, palms have always determined the vertical structure of the plant ecosystem there. The protective foliage of their wide-spreading crowns is essential for the cooler, moister microclimate in the levels below. Depending on climatic conditions and regional food preferences, other fruit-producing plants such as fig, pomegranate, jujube, orange, apricot, mango, and papaya trees often grow under the palms.

left
Hot springs in the Libyan oasis town of Ghadames, 1931.

These lower-growing plants are more sensitive to direct sunlight, heat, and dryness than the robust palms are. Grapevines
may also wind directly up these smaller trees or grow on extra
supports provided for them. Farther down at ground level, farmers often plant annuals that are useful to humans in one way or
another: vegetables, herbs, grain, cotton, tobacco, or cannabis,
for example. Combining plants in these three levels increases
the land's productivity without significantly changing the effort
needed for irrigation and remains a characteristic method in the
Middle East and North Africa today. Palm oases without such sub-
layers do not require intensive cultivation, so nomads can simply
visit them occasionally to pollinate the blossoms and harvest the
dates.

Over the centuries, travelers have recorded the impressions palms made on them, especially when they were planted in the vicinity of urban settings, where utility was combined with pleasure. After reading the account of the Austrian journalist Amand von Schweiger-Lerchenfeld (1846–1910), describing what he experienced in late-nineteenth-century Baghdad, who wouldn't want to visit a palm grove:

> *They are surrounded by dry walls whose tops appear overgrown with brush and vines, and between the shallow irrigation channels lush vegetation blooms and sprouts, shaded by the crowns of pleasant lime trees, oranges, and figs over which the stately palms tower with their bountiful bunches of fruit. Within this green, enlivening atmosphere are small dwellings whose cool lower rooms provide shelter during the unbearably hot summer days, while the evening lures the occupants to the terrace where not only the cool breeze but the swollen fruits hanging down from the treetops to these platforms are an invitation to take refreshment. Such gardens can be found directly below the city on both sides of the Tigris.*

— 3 —

Gardens
of the
Gods

WHILE A WEALTH OF differences existed among early orchards in the broader Middle East, a distinction can be made between those associated with the royal lifestyle and those intended for cultivating trees efficiently to maximize the resulting fruit harvest. Orchards of the second type were located outside of settlements, while the fruit trees included in the first type were often incorporated into large gardens intended for many purposes other than simply fruit production, including as places to rest, display wealth, and give form to intellectual inquiries, especially those that concerned the owners' current and future relationships with the gods. Impressions of the palace gardens can occasionally be gleaned from sketches, reliefs, and even paintings, as well as archaeological remains. Simpler family gardens or orchards operated as businesses—which inspired practically no aesthetic consideration at all—have generally disappeared without a trace, or at most left a few faint clues in the desert sand.

IT WAS FOR GOOD reason that the ancient historian Herodotus referred to ancient Egypt as "the gift of the Nile." Beyond the Nile delta and the narrow green shores of the river that bisects the land, agriculture was only possible with a continuous irrigation system. It was essential to reroute the river water and better allocate it to otherwise infertile areas. It's no surprise that building a system like this was expensive, so irrigated land became a sign of wealth.

Those who could afford to irrigate land for gardens were not just concerned with burnishing their reputations with their peers—as we shall see, creating a garden had implications for the afterlife, and artworks depicting gods watching over the cycle of nature acknowledged that they were the ones who ensured that the garden would be fruitful.

A Theban wall painting from the Eighteenth Dynasty (1554 to 1305 BCE) includes a tree goddess in a sycamore holding out the tree's fruit. Sycamores are deciduous trees from the mulberry family. They can grow up to nearly fifty feet (fifteen meters) high and their crowns can measure eighty feet (over twenty-five meters) around. The trees' fig-like fruit grows directly from the trunk or older branches in clusters resembling grapes—a phenomenon that botanists call by the lovely name *cauliflory*. While the sycamore's fruit does not taste nearly as good as "real" figs, it was still a delicacy. In addition to date palms and sycamores, Egyptian orchards included pomegranate trees, which were imported from lands farther east. Once Egypt had grown into a mighty empire, it imported

previous page
Orchard in Deccan, India, ca. 1685.

cultural influences from Syria, Palestine, the rest of the Mediterranean, and regions to the south.

Ancient reliefs show small groves growing around artificial pools as part of the landscaping for individual tombs. The fruit from the trees was intended to protect the deceased from thirst and nourish them, and the large palm leaves were there to fan fresh air their way. On special occasions a statue of the departed might be rowed across the pool so that they could enjoy the sight of their garden.

above
Sobekhotep's tomb
in Thebes (now Luxor,
Egypt), ca. 1400 BCE.

above
Picking and eating
figs, Khnumhotep II's
tomb at Beni Hasan,
Egypt, ca. 1950 BCE.

We can thank papyrus texts such as the *Book of the Dead*, created in 1250 BCE, for our relatively good idea of what these fruitful gardens looked like. In a vignette in chapter 58, Ani and his wife, Thuthu, are depicted in a pool with water running through it. In their left hands, they hold a sail, the symbol of air, while they dip their right hands into the water to drink. Palm trees hung with clusters of dates surround the pool.

We can also look to the work of Egyptologists from later centuries, such as the painstaking drawings of Ippolito Rosellini (1800–1843). As part of the Franco-Tuscan expedition to Abu Simbel, Rosellini copied hundreds of images from Egyptian monuments. Some show gardens from a bird's-eye perspective, while others depict scenes such as the grape or fig harvest. His most important work, *I Monumenti dell'Egitto e della Nubia* (*Monuments*

From Egypt and Nubia) (1832–1844), is a collection of many of these colorful drawings. If the surviving images are accurate, the gardens laid out in or near palaces were compact, stylized realms with symmetrical rows of trees around a pool lined with ornamental plants and harboring fish, water lilies, and birds. Sometimes lightly clad people also appear, captured in graceful poses as they care for the trees or harvest their fruit.

These palace gardens were also places where humans could make sacrifices to the gods or dedicate objects to them. Since gardens would be especially pleasant places for gods to spend their time on earth, they seem particularly suitable locations for such rites and devotional practices. The appropriate gods for a garden depended on the type of fruit cultivated there. The god of the grapevine, for example, was Osiris. The grape harvest had its own specific deity, the goddess Renenutet, and even the equipment used to press the grapes had a divine spokesperson in the lion- or ram-headed god Shesmu.

Researchers have learned that Egyptian viniculture has its roots in the prehistorical Negade period. The grapes from the northeastern delta and oases were especially prized, which tells us that vineyards were located outside the cities and palaces. The vines grew on trellises cleverly constructed of horizontal sticks and forked posts. During the Ramesside period, vineyards expanded to the point that a single one could require more than a hundred workers. In addition to grapes, wine was made from figs, dates, and pomegranates, and the texts from that period often blur the distinctions between vineyards, orchards, and palm groves. Traders were transporting wine between Egypt and the Levant (or, at that time, Canaan) at the end of the fourth millennium BCE: in Abydos, a city in central Egypt on the western bank of the Nile, a series of wine amphorae made in Palestine were found in the tomb of Pharaoh Scorpion I (ca. 3200 BCE).

The Egyptians also cultivated other types of orchards solely for the purpose of growing fruit. (Of course, the people working in these places may also have enjoyed moments of pleasure or relaxation there, but we have no documentation that they did.) Discoveries at the central Egyptian Tell el-Amarna archaeological site, on the Nile's eastern shore, indicate how such large commercial areas were divided into separate sections for fruit, vineyards, and vegetables. A wall surrounded the entire complex, and the plots for different kinds of crops were also enclosed.

Despite the vast stretches of time that have passed, we have some evidence about the work that went on at places like Tell el-Amarna. The gardeners did not enjoy high social standing and the working conditions were harsh: they toiled all day in the burning sun without even a head covering to protect them from the heat. Carrying heavy jugs of water often left them with sores on their necks. Foremen in charge of the crews drove them to keep working. But as some images from tombs show, workers occasionally had the chance to rest and recover in the shade.

As the fruit was ripening, one important task was to drive off the swarms of starlings and parakeets that happily plundered the trees and vines. (The technique of covering the date bunches with nets to protect them from hungry animals was developed later.) Baboons were notorious for ripping out blossoms and the hearts of young palms. Rats were a threat too, and swarms of locusts could destroy an entire crop in minutes.

To harvest the fruit, workers climbed into the trees with baskets on their backs, picked the fruit, and lowered the filled baskets with ropes. Ladders were apparently not commonplace, but tame monkeys were put on leashes and sent up the fig and palm trees to collect their bounty.

A sense of pleasure wafts through even the most production-focused orchard. The "Turin Erotic Papyrus," a manuscript from the time of the New Kingdom (1550–1080 BCE), contains love poems in the voices of various fruit trees. (This fascinating document also contains the oldest known description of sexual intercourse—an early example of the long association between eating fruit and other sensuous pleasures.) The pomegranate's words provide a taste:

> *Like her teeth my seeds,*
> *Like her breast my fruit,*
> *[foremost am I] of the orchard*
> *since in every season I'm around.*

DISCOVERIES IN MESOPOTAMIA ALSO point to the existence of different kinds of orchards and gardens: those that served representative and recreational purposes for the elite, and those devoted to the production of food. This distinction was similar to the one in Egypt, but conditions in the two countries were very different: in their natural state, nothing about Mesopotamia's wide, often mud-covered plains was suitable for growing plants. The people first had to contain the waters of the Tigris and Euphrates and then dig canals to channel it onto the fields. Without this extensive reshaping of the landscape, the existence of palm groves and cultivated fields—which had spread to cover the entire land by the first half of the third century BCE—would have been impossible.

Clues exist to what these first gardens might have looked like, but they all date from much later. The Assyrian king Assurnasirpal II (ninth century BCE) took great pains to link Nimrud, his capital on the Tigris in what is now northern Iraq, with sources of water:

I dug out a canal from the Upper Zab, cutting through a mountain peak, and called it Abundance Canal. I watered the meadows of the Tigris and planted orchards with all kinds of fruit trees in the vicinity. I planted seeds and plants that I had found in the countries through which I had marched and in the highlands which I had crossed: pines of different kinds, cypresses and junipers of different kinds, almonds, dates, ebony, rosewood, olive, oak, tamarisk, walnut, terebinth and ash, fir, pomegranate, pear, quince, fig, grapevine . . .

The canal-water gushes from above into the gardens; fragrance pervades the walkways, streams of water as numerous as the stars of heaven flow in the pleasure garden . . . Like a squirrel I pick fruit in the garden of delights.

In a region where few trees grew naturally, a collection like this would have been quite a luxury. What's more, since the trees came from throughout the empire and required elaborate care, they also demonstrated the ruler's power and influence. When Sargon II (eighth century BCE) built the new city of Dur-Sharrukin, he included gardens as well. Visitors could even hunt lions and practice falconry there, so these royal grounds must have mimicked natural arrangements of plants and features and been more extensive and diversely designed than their Egyptian counterparts. In these spaces, trees that provided fruit were fully integrated into grounds intended for rest and relaxation, and the wildlife these spaces attracted was as important as the plants that grew there.

Sargon II's successor, Sennacherib (seventh century BCE), had an aqueduct built to bring water from the mountains to the Assyrian city of Nineveh, which was located in Upper Mesopotamia on the eastern bank of the Tigris, near Mosul in current-day Iraq. (In the last thirty years, scientists have identified Nineveh as the probable site of the legendary Hanging Gardens of Babylon.) In the king's words:

> To arrest the flow of the water through those orchards, I made a
> swamp and set out a cane-brake in it: herons, wild boar, beasts
> of the forest I released into it. By the command of the gods, vines,
> all kinds of fruit trees and herbs thrived and burgeoned within
> the gardens. Cypress and mulberry trees grew large and multi-
> plied; cane-brakes grew fast to mighty proportions; sky-birds
> and water-birds built their nests; and the wild sows and beasts
> of the forests brought forth their young in abundance.

A later limestone relief from Nineveh shows Ashurbanipal (seventh century BCE), king of the Neo-Assyrian Empire, lounging under an arbor of vines in a garden with bird-filled trees. The queen sits next to him in a kind of elevated chair. This so-called Garden Party scene gives us an impression of graceful, sophis-ticated people and shows a wide range of plants.

Aside from the plant designs in the decorations of temples, nearly everything related to Mesopotamian gardens has fallen victim to time. However, archaeologists assume that, as in Egypt, there were plots of cultivated land outside the cities surrounded by high walls made of clay brick, packed earth, or stones as pro-tection from thieves and animals. The walls marked a boundary showing that the plants inside them had been selected and com-bined into a new ensemble that followed different rules than the outside world. They contained a treasure that required care and protection so that it could ripen into something benefit-ing humans. They also symbolized a claim of ownership. And depending on how well they screened the garden from its surroundings—denying anything more than a glimpse of what

was behind the garden's wall—they may well have awakened longings and fueled fantasies of the treasures inside that perhaps bore little resemblance to reality.

Low-lying areas at the level of the river or even below the irrigation canals were most suitable for growing trees and other plants. But the *shaduf*—a lever-operated well—was soon widely used to bring water to somewhat higher elevations. The city was not an ideal place for productive gardens and orchards for the simple fact that new structures were continuously built on the ruins of the old ones: irrigation water would have to be brought up to increasingly higher elevations.

One key technical innovation in this area was the device known as Archimedes' screw, which made it much easier to pump water upward from cisterns which, in turn, were fed by canals. (Interestingly, this type of pump seems to have been in use four hundred years before Archimedes was born; it probably gained its name because Archimedes was the first to describe how it works.)

In Akkadian, one of the kingdom's official languages, the word for orchard is *kiru*. The gardeners were called *nukarribu*. Their job required more than just knowledge of trees: they also had to understand the irrigation system and give instructions for creating an entire network of channels so that every single plant would receive the water it needed. The names preserved in written sources, such as "Palm Grove on the Bank of the Tigris," provide further evidence that these orchards could be found alongside rivers or irrigation canals. Most gardens were administered by the government and were either run by palace officials or leased. In the latter case, operators paid a fee based on the harvest as estimated by a palace inspector—a system intended to prevent officials from attempting to defraud the government or disputing the fee amount.

But Mesopotamian orchards were more than just places where people—using the knowledge available to them—cultivated plants in order to harvest precious fruit. They were also one of the few spots where people outside their homes could be comfortably protected, at least to some degree, from the sun. A group of intentionally planted trees creates a microcosm in which life follows a predictable cycle. Evaporation from irrigation systems and ponds cools the air. The shade of the trees and the luscious fruit to be found there produce a refreshing contrast to the world that the people face outside the garden walls. In addition to the fruit trees, poplars and tamarisks were often planted along canals and waterways to prevent soil erosion.

To see just how developed and sophisticated the management of these early gardens was, we can look to one of the world's earliest surviving orchards: the Agdal, an enormous property south of Marrakesh that belongs to the Moroccan royal family. The name comes from the Berber language and means "walled meadow." Abd al-Mu'min built the nearly two-square-mile (five-hundred-hectare) Agdal in the twelfth century, but many changes have been made to it in the intervening years. Today a nineteenth-century wall topped with numerous small towers surrounds its countless trees: date palms, olive trees, and trees bearing figs, almonds, apricots, oranges, and pomegranates.

Underground channels supply fresh water for the Agdal and drinking water for the neighboring city. These channels originate in the Ourika valley, nearly twenty miles (thirty kilometers) away

in the mountains of the High Atlas, which form a dramatic panoramic backdrop to the orchard. A series of pavilions are strewn across the grounds, which also contain three rectangular pools. In the past the largest was used to teach soldiers to swim. Now carp frolic in its waters, and the Agdal is a UNESCO World Heritage site.

In the vicinity of the Agdal, traditional fruit orchards of a different type can be found. These remarkable groves—which grow wild in some cases—primarily exist today on the Atlantic coast in the southwest part of the country, somewhat west of Marrakesh and south of Essaouira. They consist of argans, gnarled trees that can grow up to forty feet (twelve meters) tall. Argans could once be found throughout vast stretches of the Mediterranean region and North Africa. They are relics from a time before the massive climate shift to our current conditions took place, when tropical

above
The walls of the Agdal, south of Marrakesh, Morocco.

conditions were the norm all the way to the poles. In fact, the argan is often considered a "living fossil" from this earlier phase of the earth's history. In the present day its range is restricted to south-western Morocco, between the Atlas Mountains, the Atlantic Ocean, and the Sahara. Olive trees also grow there, but argans are easy to distinguish from them based on their bark, which resembles snake-skin. Their small yellowish berries fall from the trees when they are ripe and can be gathered by hand. They contain hard, oily seeds known as argan nuts.

With roots that can extend up to a hundred feet (thirty meters) below the ground, argans play an important role in preventing desertification. They also have the unusual ability to survive droughts by shedding their leaves and fall-ing into a kind of "sleep." Argans only reach their full fruit-bearing capacity between the ages of fifty and sixty. Berber women traditionally cracked the seeds with river rocks so that they could process them. The nutty-tasting oil is a popular accompaniment to bread or couscous and is also used as a hair treatment and cooking fat. The trees' hard wood is also valuable for many purposes. Goats love not only argan leaves but the fruit: they clamber into the trees and eat the berries, and the seeds later come out the other end. People took advantage of this natural process by collecting these "pre-processed" seeds already freed from their fruity covering.

But let's leave present-day Morocco and return to the region of our Mesopotamian gardens, which in the course of the following centuries has become a primarily Muslim land. Accounts of the gardens of Persia make them seem like manifestations of people's longing for paradise. The 214th story of the fabled *Arabian Nights* provides an example that underscores the interweaving of pleasure grounds and productive spaces in these times:

> They entered through a vaulted gateway that looked like a gateway in Paradise and passed through a bower of trellised boughs overhung with vines bearing grapes of various colors, the red like rubies, the black like Abyssinian faces, and the white, which hung between the red and the black, like pearls between red coral and black fish. Then they found themselves in the garden, and what a garden! There they saw all manner of things, "in singles and in pairs." The birds sang all kinds of songs: the nightingale warbled with touching sweetness, the pigeon cooed plaintively, the thrush sang with a human voice, the lark answered the ringdove with harmonious strains, and the turtledove filled the air with melodies. The trees were laden with all manner of ripe fruits: pomegranates, sweet, sour, and sour-sweet; apples, sweet and wild; and Hebron plums as sweet as wine, whose color no eyes have seen and whose flavor no tongue can describe.

left
Goats in argan trees, Morocco.

When writer and passionate gardener Vita Sackville-West traveled through Persia in the twentieth century, she found the gardens—although many of them had since been neglected—no less enchanting than the one described all those centuries before, and fully understood that they represented something completely different from what she knew from her home in England. In her 1926 book *Passenger to Teheran*, she wrote:

> But they are gardens of trees, not of flowers; green wilder-nesses... Such gardens there are; many of them abandoned, and these one may share with the cricket and the tortoise, undis-turbed through the hours of the long afternoon. In such a one I write. It lies on a southward slope, at the foot of the snowy Elburz, looking over the plain. It is a tangle of briars and grey sage, and here and there a judas tree in full flower stains the whiteness of the tall planes with its incredible magenta. A cloud of pink, down in a dip, betrays the peach trees in blossom. Water flows everywhere, either in little wild runnels, or guided into a straight channel paved with blue tiles, which pours down the slope into a broken fountain between four cypresses. There, too, is the little pavilion, ruined, like everything else; the tiles of the façade have fallen out and lie smashed upon the terrace; people have built, but, seemingly, never repaired; they have built, and gone away, leaving nature to turn their handiwork into this melancholy beauty... The garden is a place of spiritual reprieve, as well as a place of shadows. The plains are lonely, the garden is inhabited; not by men, but by birds and beasts and lowly flowers; by hoopoes, crying "Who? Who?" among the branches; by lizards rustling like dry leaves; by the tiny sea-green iris.

Of course, the idea that paradise was a fruitful garden is an old one found in many religions. The account from the Bible reads:

> *The Lord God took the man and put him in the Garden of Eden*
> *to work it and take care of it. And the Lord God commanded the*
> *man, "You are free to eat from any tree in the garden; but you*
> *must not eat from the tree of the knowledge of good and evil, for*
> *when you eat from it you will certainly die."*

But Adam and Eve gave in to temptation and ate the fruit of the one proscribed tree, and humanity was forever banned from paradise. In their shame, they hid from God under a canopy of leaves. What kind of fruit was responsible for this terrible turn of events? It was certainly not an apple, despite the many pictures of the story showing Eve with one. In fact, there is no mention of an apple at any point in the creation story—the only term used is "fruit."

The persistent belief in Eve and the apple was most likely born from an error or even an intentional play on words by the early translators who created the vernacular Latin version of the Bible. They confused the words *malus,* which means "apple," and *malum,* which means "evil." But the phrase *lignumque scientiae boni et mali* simply means "the tree of the knowledge of good and evil," with nary an apple to be found. It was just a small step to the first illustrations of Eve with an apple in her hand. And since the fifth century CE this misunderstanding has persisted down through the centuries to today.

So if it wasn't an apple, what was it? An apricot, a fig, a pomegranate, a date? We may scour the Bible all we want, but the identity of the forbidden fruit will remain a mystery.

— 4 —

Not Far From the Tree

RESEARCHERS KNOW MORE ABOUT the history of the apple than any other type of fruit. Primitive forerunners led to the wild-growing apples found today in North America (*Malus coronaria, Malus fusca*), East Asia (*Malus sargentii, Malus sieboldii*), China (*Malus hupehensis, Malus baccata*), and in the Himalayas (*Malus sieversii, Malus orientalis*), all of which are known as crabapples because of their small size and sour taste. It can be difficult to tell the difference between crabapples and some cultivated varieties that have gone wild. Scientists disagree, for example, about whether *Malus sylvestris*—the European crabapple—is in fact wild or actually a cultivated apple that more or less resembles its wild cousins. All these wild apples hail from the temperate zone

of the Northern Hemisphere, where the cold period important for seed growth takes place.

Wild apples are small and most are bitter, but when they are dried, they definitely have value because the flavor of the fruit intensifies when you dehydrate it. Archaeologists have discovered traces of dried wild apples that are about 4,500 years old at the sites of human settlements near Swiss lakes and, threaded into chains, in the tomb of Queen Puabi in the Mesopotamian city of Ur. Wild apples are on the menu for many animals, including bison, deer, bears, wild boars, and badgers. And thanks to their dense network of branches, wild apple trees are attractive hiding places for small animals. Owls raise their chicks in the hollow trunks, while bats can retreat there in the daytime.

Since crabapples are just an inch or two (or a few centimeters) in size, birds—which are especially attracted to small hanging fruit—probably played a central role in the trees' spread. Horses show a pronounced fondness for apples, especially those on the sweeter side, and researchers believe it is possible that the horses of nomadic traders are responsible for isolated populations of small sweet apples in the Caucasus and Crimea; in parts of Afghanistan, Iran, and Turkey; and in the European section of Russia near what is now the city of Kursk.

Our cultivated apple (*Malus domestica*) with its many varieties is the result of cross-breeding among the kinds of wild apples that still grow in central Asia today. Scientists have determined that this region, with its alternating semi-arid areas, plateaus, and mountainous landscapes, is home to an especially dense variety of plant forms. Environmental conditions vary greatly there over a small space, and researchers believe that this variety was important for the processes determining early fruit development.

It is likely that the fruit forests on the slopes of the Tian Shan mountains played a central role in this evolutionary drama. These tall mountains (the peak of Jengish Chokusu, for example, is over 24,000 feet or nearly 7,500 meters high) stretch over a thousand miles (sixteen hundred kilometers) between western China and Uzbekistan. Many of the fruit species typical for the Northern Hemisphere's temperate zones can be found here, from apples to pears, quinces, apricots, cherries, plums, cranberries, raspberries, grapes, and strawberries to almonds, pistachios, hazelnuts, and walnuts.

The area provided all the required climatic and ecological conditions, including a varied landscape with diverse plant life and more than two hundred rivers, combined with long periods of time in which the land remained untouched from outside—deserts sealed it off in nearly every direction. The apple trees there come in different shapes and sizes, and the fruit offers a wide range of flavors: some sweet with a taste of honey, anise, or nuts, and others so sour that they are nearly inedible. The early sweet wild apple, the original ancestor of all our cultivated apples, probably traveled from the location in China where it originated to Tian Shan several million years ago.

Thanks to the secluded geographical location and, until relatively recently, political isolation that has shielded the area from attention, the surviving pockets of Tian Shan's astounding forests have become a mecca for botanists seeking to study the wild plants and preserve the stores of genetic material they contain. One such researcher is John Palmer, fourth Earl of Selborne, who has managed a fruit farm on his estate in Hampshire for almost half a century and is a past chair of the trustees of the Royal Botanic Gardens in Kew. To learn more about how apples grow in the wild, he traveled to the Dzungar Alatau area in Kazakhstan, near the border with China:

> We saw, for the first time, woods where apple was the dominant species, together with poplars and maples. Hops grew up many of the apple trees as well; in Hampshire, I used to grow hops in addition to apples, and never expected to see the two plants growing in the wild together. If I had only followed nature as demonstrated at Topolyevka, I could have grown my hops up old Bramley apple trees and saved a great deal in costs by forgoing the posts and wire work I used to train the hops. Few of the apple trees were cropping because of late snow in May. Used as I am to well-pruned orchards with evenly spaced trees, my introduction to an apple forest was something of a culture shock. The

trees were densely interlocked, with branches broken by bears
harvesting the crop. Much of the regrowth came from suckers
thrown up by old trees, which formed an impenetrable tangle.

Although this earthly Garden of Eden grew without the help
of any human hand and still lies far from any human habitation,
it and the treasure trove of genetic information it contains are in
danger. Pollen from cultivated apples is one threat, while grazing
wild horses that eat the saplings is another.

How did the fruit of this forest continue to spread throughout
Asia and toward the west? The summer and fall months in central Asia's foothills and plains must have been agreeable for early
humans. The routes followed by nomads on horseback through
Asia's interior are at least a thousand years old. Later, the fruit surely
traveled with merchants along the antique Silk Road, the legendary trading route. Along with their loads of silk and ceramics, the
caravans moving from east to west also carried food, and not all of
it was intended for the travelers' own use during the long journey.

One of the places where interesting clues have come to light is
Tashbulak, a settlement established by nomads in northwestern
Uzbekistan at the edge of the Pamir mountains, about 7,220 feet
(2,200 meters) above sea level. The walnut kernels discovered during the excavation of a medieval dump site came from nearby trees,
but the remains of grapes and peaches found there came from
fruit that grew in a milder climate—perhaps near the oasis cities
of Samarkand or Bukhara to the west.

The Spaniard Ruy González de Clavijo visited this region a century after Marco Polo but was still one of the first Europeans known
to have traveled there. At the end of August 1404 he glimpsed a
large orchard resembling a park in the outskirts of Samarkand:

We found it to be enclosed by a high wall which in its circuit may measure a full league, and within it full of fruit trees of all kinds, save only limes and citron trees which we noticed to be lacking... Further there are here six large tanks for a great stream of water flows from one end of the orchard to the other. Five rows of very tall and shady trees have been planted beside the paved avenues which connect the pools and smaller paths lead out of these avenues to add variety to the design.

Beyond this orchard was another garden of similar size dedicated to vineyards.

APPLES ARE A NEARLY perfect fruit: they can be stored for longer periods than many other species and transported across wide distances. People probably discovered early that dried apple slices were even easier to keep because they no longer attracted insects or provided a foothold for the bacteria or molds that promote decomposition. Drying also reduces the bitter taste that some apples have. In the Middle Ages, crabapples were primarily used as seasoning for other foods. And carpenters and woodworkers prized this tree's hard, usually spiral-grained wood, which they used for clock hands, treadmills, screws, and furniture veneers.

The apple doesn't fall far from the tree. A well-known saying, but let's take a step back and consider it literally: it refers to the fact that as soon as an apple is ripe it falls to the ground. This cause-and-effect reaction depends on communication between the tree and its fruit. Apples produce ethylene as they ripen. When the tree receives this signal, its leaves begin to produce abscisic acid, a developmental hormone. A barrier layer forms between the twig and apple stem as a result, cutting off the flow of nutrients and causing the fruit to fall off.

right
The Swedish artist Carl Larsson offers a view into an apple orchard at harvest time, early twentieth century.

We might further imagine that this apple gradually decomposes and one of the seeds finds its way into the ground at this spot. However, the outlook for the resulting sapling is not especially rosy: its parent would block the sun and compete with it for water and nutrients. To get the sun and air they need to thrive, apple trees must be separated by a certain amount of space. This is where birds and animals come in to help the apples spread their seed. Interestingly, young apple seeds are resistant to cold thanks to natural germination inhibitors they contain. They need to pass through a winter and first sprout in the following vegetation

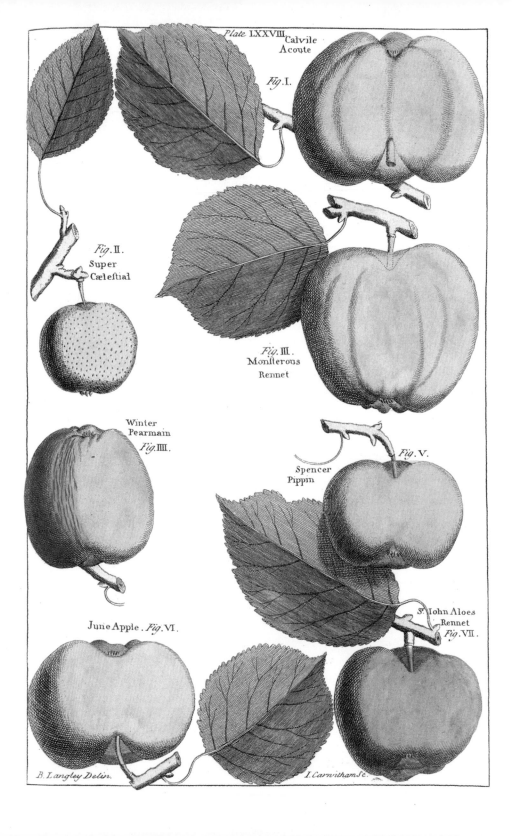

Plate LXXVIII. Calvile
Acoute

Fig. I.

Fig. II.
Super
Cælestial

Fig. III.
Monsterous
Rennet

Winter
Pearmain
Fig. IIII.

Spencer
Pippin

Fig. V.

St. Iohn Aloes
Rennet
Fig. VII.

June Apple. Fig. VI.

B. Langley Delin.

I. Carwitham Sc.

period, after they have absorbed moisture and swelled. If this were not the case, the young plants would freeze during the cold part of the year.

Quite apart from the issue of seed dispersal, the idea that an apple "doesn't fall far from the tree" contains a basic misunderstanding concerning apple genetics. While human children—in line with the meaning of the expression—are indeed sometimes very similar to their parents, apple seeds are not. Each seed contains multiple versions of genetic material, and the oldest are generally the most dominant. As a result, most apple trees that grow from seeds don't even produce fruit that is edible for humans. Such an offspring actually falls "far from the tree" that produced it.

So it's very rare that a fruit tree grown from a seed will have exactly the same characteristics as its parent. In fact, producing a series of plants with the same genetic characteristics requires circumventing the natural process of sexual reproduction. Making copies of trees or fruit that are especially attractive or good-tasting requires human intervention in the form of grafting. In this method, the shoot of a particularly desirable tree, known as a scion, is spliced with a rootstock, the root system and part of the trunk of another tree. For a graft to be successful, the scion's cambium—a layer of growth below the bark—must be positioned on top of the cambium of the rootstock so that an intermediary tissue forms and the two elements can fuse. This process takes two to four weeks.

If everything is done properly and there are no incompatibilities between the scion and the rootstock, the two parts grow into one and become a copy—a natural clone—of the tree from which the scion was taken. The rootstock ensures that the new tree receives water and nutrients and also influences its growth in a variety of ways. Grafting also speeds up the tree's developmental

left
Apples from Batty Langley's *Pomona*, 1729, a publication that promised "sure methods for improving all the best kinds of fruits now extant in England."

clock: if the scion comes from an older tree, its mature blossoming behavior is transferred to the new plant no matter what the age of the rootstock may be. The result is a tree that produces blossoms and fruit earlier than it normally would.

Grafting ushered in a whole new level of fruit tree cultivation. The ability to produce fruit varieties whose offspring always have identical characteristics depends on it. Over millennia, people have steadily learned how to improve this technique.

The precise origins of grafting are a mystery. Barrie E. Juniper and David J. Mabberley, who have intensively studied the apple's historical development, formulated an interesting theory that farmers working in the fields may have observed an unplanned version of the grafting process. If parts of neighboring plants were woven into a mat, for example, it could have been possible for these parts to combine and for new shoots to spontaneously develop from them. Observing this phenomenon could have motivated people to intentionally create such combinations. Observations like these could have occurred in different places and the technique transferred to other types of fruit trees over time.

Determining which plants are suitable for grafting and which are not is quite complicated and influenced by a number of factors. Over time, people gained experience in which pairings of fruit species are compatible. One basic rule is that the plants must be relatively near relations. As the number of genetic similarities between two plants increases, so does the likelihood that their fusion will be successful.

The rootstock determines the height of the grafted tree, while the scion determines the type of fruit it bears. Today pear scions are grafted to quince stocks, but apple trees are not amenable to such cross-species grafting. Almonds and peaches are compatible, but almonds and apricots are not—even though all three belong to the genus *Prunus*. Showing that he already recognized these principles, Theophrastus wrote: "Like always coalesces readily with like, and the bud is, as it were, of the same variety." But past ideas about which plants could be grafted together often reached fantastical heights. For example, one theory held that grafting mulberry branches to a white poplar would create a tree that produces white mulberries.

In addition to the biological laws governing the behavior of grafted plants, cultural and religious attitudes became attached to grafting, which was perceived as akin to a marriage. In this way of thinking, a combination of unequal plants, such as a cultivated apple and a wild pear tree, should be avoided. After all, the reasoning went, the child of an educated family should not be married to an illiterate barbarian.

Over time, the share of wild fruit-bearing plants fell as more fruit trees were intentionally planted. As a result, cultivated fruit varieties emerged that increasingly differed from their wild counterparts and reflected the desires of gardeners and farmers. Patience was a necessary virtue: at least three years must pass after a tree is planted before it bears fruit for the first time. Furthermore, young trees require especially careful tending.

Productive trees that lie somewhere on the broad spectrum between "wild" and "cultivated" still exist today. For example, people in central Anatolia intentionally preserve wild fruit trees: primarily pears, but also the Cornelian cherry, Mount Atlas mastic, almond, and hackberry. They simply enjoy the resulting bounty or sometimes graft shoots of older trees with especially tasty fruit to the rootstocks of young plants. Twigs of cultivated olive trees are also sometimes grafted to oleasters, their wild counterparts.

— 5 —

Studying the Classics

PEOPLE IN ANCIENT GREECE were familiar with the technique of grafting. They were always in close contact with their neighbors in Asia Minor, which was their source of both fruit trees themselves and methods for cultivating them. Homer provides the first known description of an orchard—specifically, the one belonging to Alcinous, the king of the Phaeacians—in the *Odyssey*, Book VII, written around the turn from the eighth to the seventh century BCE. In the majestic translation by Alexander Pope, it reads:

> *Close to the gates a spacious garden lies,*
> *From the storms defended and inclement skies;*
> *Four acres was the allotted space of ground,*
> *Fenc'd with a green enclosure all around.*

Tall thriving trees confessed the fruitful mold:
The red'ning apple ripens here to gold,
Here the blue fig with luscious juice overflows,
With deeper red the full pomegranate glows,
The branch here bends beneath the weighty pear,
And verdant olives flourish round the year. ...
Each dropping pear a following pear supplies,
On apples apples, figs on figs arise.

At another point, the *Odyssey* describes the homestead belonging to Laertes, Odysseus's father. In addition to a house, dwellings for slaves, stables for animals, fields, and vineyards, it lists the same types of fruit as those found in the garden of Alcinous.

The Greek word for "orchard" is sometimes the familiar-looking *orchatos* and sometimes *kêpos*, a word whose meaning is difficult to pin down precisely, somewhat like the ever-changing qualities of the garden it describes. The archaeologist Margaret Helen Hilditch has analyzed the Greek concept of *kêpos*:

The three essential "resonances" were: of care for something highly valued; of luxury, privilege and eastern elements; and of the tempting yet risky presence of women. These, combined with the Greek landscape, made the garden an ambivalent, borderline space.

Orchards outside the cities were located on rivers and streams that ensured a supply of water. The Greek poet Sappho, for instance, who lived around 600 BCE, writes of an area dedicated to Aphrodite with apple trees and meadows, as well as a bubbling spring that apparently provided irrigation.

Homer's mention of fig cultivation so early in Greece is interesting since these trees are believed to have originated in Persia.

previous page
Detail of a funerary monument for an athlete from Boiotia, Greece, ca. 550 BCE, showing two pomegranates in the upper right.

Their pollination ecology is complicated, and in most cases only functions in conjunction with the fig wasp. In addition to male and female blossoms, fig trees have gall blossoms that serve as breeding grounds for these insects. But figs with fig wasps in them are considered inedible, so orchardists bred varieties of fig trees producing only gall blossoms and male blossoms (caprifigs) and others producing only female blossoms. In a process known as "caprification," twigs from the caprifig are hung in the crowns of cultivated fig trees to pollinate them. Then these tame female trees bear edible fruit.

The Greeks also had a great appreciation for wine grapes, which reached their land from the region around the Black Sea. According to Greek mythology, Dionysus, the son of Zeus, discovered them and introduced them to humankind. The great philosopher Plato was of the opinion that "nothing more excellent or valuable was ever granted by God to man."

The oldest known seeds of wine grapes—which presumably were already being cultivated—were discovered in the Transcaucasia, in present-day Georgia. They date from the sixth millennium BCE. Early wine growers specifically selected vines that could pollinate themselves and did not require interaction with another plant so the grapes they grew would always breed true. To this day, most varieties of grapes used for wine-making are self-fertile. Evidence from six thousand years ago indicates that the technique of fermenting grape juice into wine was invented in the western Caucasus as well. Today Georgians still produce wine using methods that have existed for thousands of years. This includes burying earthenware jugs called *kvevri* for five to six months to age the wine—essentially an example of living archaeology.

It seems clear that Greek historians were well informed about this part of the world early on. Herodotus wrote: "Many and all manner of nations dwell in the Caucasus and most of them live

——
right
Theophrastus is, for
many people, the
father of botany for
his works on plants.

on fruits of the wild wood." And he knew that the people living
east of the Caspian Sea had methods for stretching the stores of
fruit they harvested over the entire year: "In winter they live on
fruit they get from trees and store, when it is ripe, for food." He is
probably referring to plums, which were known for being eaten in
both fresh and dried forms.

One especially important figure for fruit cultivation in ancient
Greece was Theophrastus (ca. 371–287 BCE), who went down in
history as the "father of botany" and played a role in shaping the
practice of this discipline into at least the seventeenth century.
Legend has it that he held his lectures while wandering through
the pathways of his garden, where he raised a variety of trees and
plants as in a botanical collection. In his two major botanical

treatises, *Enquiry Into Plants* and *On the Causes of Plants*—which have only survived in incomplete form—he mentions six types of apples, four types of pears, and two types of almonds and quince. His work contains the very first description of round, red fruits that grow on a tall tree: the cherry. Numerous varieties of wine grapes must already have existed, for he writes: "so many fields, so many types." Fruit varieties were reproduced through grafting or, in the case of self-pollinating fruit such as peaches, via seed pits. Theophrastus also mentions the technique for artificially pollinating date palms that was already long known in the Orient at that time. He refers to the peach as a Persian fruit, possibly because Alexander the Great brought peaches to Greece from Persia. In fact, we now know that *Prunus persica* originated in northern China, where it was first cultivated in about 2000 BCE. Only later did travelers on the Silk Road bring peach pits with them to Persia and Kashmir.

People have always been tempted to pluck someone else's fruit, break off branches, or even cut down entire trees for firewood. During hostilities between different groups, trees were often vandalized on purpose. Thanks to the laws codified by Draco (ca. 620 BCE), a legislator in ancient Athens, we know that trees were considered so important to the well-being of the populace that the theft and destruction of trees could even be punished with death.

THE ROMANS EXPANDED ON this Greek knowledge base. In early times, most of the fruit on Roman tables came from trees and shrubs growing in the wild. According to legend, the wolf who rescued the abandoned twins Romulus and Remus, who would later found Rome, nursed them under a fig tree. And not just any fig tree, but the *Ficus Ruminalis* located at the Roman Forum. Apple trees stood under the protection of their own special goddess, Pomona.

The god Vertumnus fell in love with her and watched over the changing seasons, with special concern for their role in the well-being of fruit trees. (His name comes from the Latin word *vertere*, for change or turn.) Vertumnus was not only young and attractive, but also capable of changing shape—just like seeds that transform into fruit.

The oldest Latin farming handbook, *On Agriculture* (*De Agricultura*), was authored by Marcus Porcius Cato, also known as Cato the Elder (234–149 BCE). He considered grapevines the most important crop, followed by olives, figs, pears, apples, almonds, hazelnuts, and others. The second book of the *Georgics*, written by Vergil (70–19 BCE) shortly before the dawn of the Common Era, examines many aspects of tree cultivation. A systematic approach was necessary because the supply of acorns and berries in the sacred forests had run out—presumably as more and more people were collecting food and driving in their pigs to forage there. Saplings in particular required protection. Once the trunks were grown and the branches began to stretch toward the sky, the trees could be left to their own devices: "No less, meanwhile, does every wood grow heavy with fruit, and the birds' wild haunts blush with crimson berries."

Among the many plant lovers in ancient Rome, Lucius Junius Moderatus Columella (4–70 CE) has a place of honor. During his lifetime, ancient Roman fruit cultivation reached its zenith. In his book *De Re Rustica*, also translated into English as *On Agriculture*, Columella lists the factors that enabled agriculture and fruit cultivation to flourish: continuous ownership of small to medium-sized holdings, agricultural specialization and investment, growing demand from urban consumers, and agriculture-friendly policies and administration under Emperor Augustus.

left
Fresco in the Villa di Livia, Rome, creating the illusion of a garden, first century CE.

Columella provides very precise instructions for organizing orchards, with tips on the best times for planting fig, almond, chestnut, and pomegranate trees. Pear trees get thorough consideration as well:

Plant the pear tree in the autumn before winter comes, so that at least twenty-five days remain before the mid-winter. In order that the tree may be fruitful when it has come to maturity, trench deeply round it and split the trunk close to the very root and into the fissure insert a wedge of pitch-pine and leave it there; then, when the loosened soil has been filled in, throw

*ashes over the ground. We must take care
to plant our orchards with the most excellent
pear-trees that we can find.*

While some of this, at least, sounds convincing,
we really have no idea whether these methods
actually worked. In any case, the work contains pro-
nouncements on grafting that we now know to be wrong:

> *Any kind of scion can be grafted on any tree, if it is not dissimilar
> in respect of bark to the tree in which it is grafted; indeed if it
> also bears similar fruit and at the same season, it can perfectly
> well be grafted without any scruple.*

This misleading assurance is offered in conjunction with
astrological tips, such as a suggestion to do any required grafting
during the waxing moon.

On terms of thoroughness, Pliny the Elder (23–79 CE) out-
stripped his distinguished forebears. His thirty-seven-volume
Natural History describes a thousand plants, including thirty-
nine types of pears, twenty-three types of apples, nine types
of plums, seven types of cherries, five types of peaches, and six
types of walnuts. But the trophy for the best-represented fruit of
all—seventy-one varieties—goes to wine grapes. While Columella
covers only the types of fruit known in Italy, Pliny looks beyond
his immediate surroundings to include varieties from Greece,
Gaul, the Orient, and Spain. He even claims the existence of a seedless apple variety—a dream that many orchardists have pursued
over time, and that has only recently been realized.

Traditional Roman gardens were planted during the period
from the sixth century BCE to the fifth century CE. They reflected

the climate conditions wherever they were located, whether on the Italian peninsula or in the conquered provinces of North Africa, Asia Minor, the Iberian Peninsula, or central and western Europe. Likewise, as the Roman Empire expanded, new plants such as date palms, pomegranate trees, and plum and cherry trees found their way to the Eternal City and other locations throughout Roman territory.

Orchards were planted near the cities so that the fruit could quickly be brought to the marketplaces there. Dealers could apparently bid on the bounty while it was still hanging from the branches. In Pliny the Elder's account, money practically does grow on trees:

> In the matter of fruit-trees no less marvellous are many of those in the districts surrounding the city, the produce of which in every year is knocked down to bids of 2000 sesterces per tree, a single tree yielding a larger return than farms used to do in the old days.

Depictions of gardens containing fruit trees painted on the walls of homes in Rome itself and the cities destroyed by Vesuvius can give us a sense of what these gardens were like. One example is the fresco from the Garden Room in Rome's Villa of Livia. This work, which dates from about 30 CE, shows pomegranate shrubs bearing fruit, some of which has burst open. Equally impressive are the frescos in Pompeii's Casa dei Cubicoli floreali, also known as the Casa del Frutteto or House of the Orchard. One scene shows a garden landscape with oleander, myrtle, lemon, and cherry trees. In another, the careful observer can spot a fig tree and a snake slithering up the trunk, apparently in pursuit of the ripe fruit. The scenes also show fences, vases, and sculptures, and the remains of identical items were found during excavations at the site. It therefore seems that the frescos were intended to make the small garden appear larger and create the illusion of a beautiful view—not unlike a modern photo mural. Because the building contained approximately one hundred amphorae, researchers believed that the last resident before Vesuvius erupted in 79 CE was a wine merchant.

The research of American archaeologist Wilhelmina F. Jashemski conveys a good idea of what the gardens of Pompeii's inhabitants were like. Most houses had at least one garden, and some had three or even four large peristyle gardens, which were gardens enclosed and surrounded by colonnaded walkways. The older houses in particular often had gardens in the back that were planted with vegetables and trees bearing fruit, olives, and nuts, along with a few grapevines. There is no evidence that Pompeiians saw a distinction between ornamental gardens and those that produced food—as in Egypt and Mesopotamia, the boundaries between these two types were fluid.

Jashemski further explains that the pleasant weather made it possible for the city's inhabitants to spend much of their time in the gardens next to their homes. Out in their gardens, they ate meals in dining spaces known as triclinia while reclining on masonry couches and enjoying the shade of encircling vines, they performed tasks such as weaving wool, or they simply relaxed. A sundial helped residents keep track of the time. Sculptures emphasized the divine aspect of nature: the garden belonging to the House of the Ephebe, for example, contained a small shrine with a statue of Pomona. Water flowed from a fruit-filled shell in her hands into a pool below. Clearly, the designers of these gardens

were alert to opportunities to honor the gods. The remains of bones
also show that a more prosaic form of protection—watchdogs—
were present as well. The birds depicted in the frescos, including
purple swamphens, herons, and egrets, surely visited the gardens
in real life. And use of the outdoor spaces did not cease when the
sun went down. On a warm summer evening, fresh garden air
was more appealing than the stuffy atmosphere inside. Evidence
shows that some of the gardens were illuminated by candela-
bras at night.

The slopes of Vesuvius near Pompeii and Herculaneum housed not only many orchards and nurseries but markets where the fruit grown in the area was sold. Jashemski describes a garden to the west of the Garden of the Fugitives in Pompeii. Part of the garden was apparently a commercial operation: the archaeologists were able to determine that the plants grew in rows, mostly on a north-south axis.

North of this garden with its mix of trees and other plantings is a cistern that presumably collected rainwater channeled from the roof of the house. A compacted depression leading through the garden to a dining area shaded by the branches of three large trees could have served either as a walking path or, perhaps, as an irrigation canal. Even a stonework altar was unearthed—archaeological evidence that sacrifices took place here, probably burnt offerings and offerings of blood. Bones and shells found at the site (from cows, pigs, goats, sheep, and snails) indicate the kinds of meals people enjoyed in this garden before it was buried in the eruption. The gods also received their share in the form of entrails burnt on the altar. Gardeners spread organic material throughout the garden to ensure the earth was fertile and mixed in moisture-storing pumice to keep the plants from drying out.

Pliny the Younger, a lawyer and senator, provides us with remarkable glimpses into the gardens of ancient Rome, where fruit trees were also part of the ornamental arrangement. He regularly took a break from his professional and official responsibilities by visiting one of his several homes outside the city. One such retreat was a *villa rustica* or country estate on the coast in Laurentum, near the port of Ostia. Of course, no country estate is complete without some provision for food production, so it is no surprise that vegetables, fruit, grapes, and olives were grown there. A letter Pliny wrote to his friend Gallus describes the house, which certainly sounds delightful:

Next to the gestatio *[exit or driveway] and running along inside it, is a shady vine-plantation, the path of which is so soft and easy to the tread that you may walk bare-foot upon it. The garden is chiefly planted with fig and mulberry trees, to which this soil is as favourable as it is averse from all others. Here is a dining-room, which, though it stands away from the sea, enjoys the garden view which is just as pleasant: two apartments run round the back part of it, the windows of which look out upon the entrance of the villa, and into a fine kitchen-garden.*

below
Pliny the Younger often retreated to his garden near the coast of the Tyrrhenian Sea.

The grounds of homes like these were designed to be used in any type of weather and to provide protection from the wind. After all, stressed-out members of the Roman elite, like Pliny, appreciated the chance to remove their togas so they could enjoy the surroundings unencumbered.

One of Pliny's other villas was located in Tuscany by the town of Tusculum. In another letter—this time to his fellow senator Domitius Apollinaris—Pliny described the interplay of different elements in the garden there. In some cases, nature had been stylized into works of art, while in other cases it retained its irregular, unplanned quality:

In one place you have a little meadow, in another the box [hedge]
is cut into a thousand different forms, sometimes into letters,
expressing the name of the master, sometimes that of the artist;
whilst here and there little obelisks rise, intermixed alternately
with fruit trees; when on a sudden in the midst of this elegant
regularity you are surprised with an imitation of the negligent
beauties of rural nature.

What happened to the fruit that all these gardens and orchards produced? Some was eaten fresh, while some was processed. Grapes were turned into wine, nuts were dried, and, as the multitalented scholar Marcus Terentius Varro (116–27 BCE) writes, the methods for storing apples were already quite advanced:

It is thought that all [apples] keep well in a dry and cool place,
laid on straw. For this reason those who build fruit-houses are
careful to let them have windows facing north and open to the
wind; but they have shutters to keep the fruit from shrivelling
after losing its juice, when the wind blows steadily. And it is for
this reason too—to make it cooler—that they coat the ceilings,
walls and floors with marble cement.

Olives, too, were a staple that was carefully harvested and processed. The practice of knocking down the fruit with long poles and gathering it in baskets was already an old one. Only a small portion of the harvest was eaten in fruit form. Most of the olives were brought to a press, where a rotating stone crushed them to extract their oil. This precious liquid was channeled directly from the press into a cistern in the ground, and from there people could fill clay jugs to take home.

THE ROMAN EMPIRE EVENTUALLY stretched over central and northwestern Europe. To understand the conditions that the Romans found there, it is helpful to look back at what these regions were like before they arrived.

Excavations of stilt-house villages from the Neolithic period (3500–2200 BCE) on Lake Constance and throughout northern Switzerland and Upper Austria show that apples, pears, plums, and sweet cherries all existed, along with hazelnuts and the occasional walnut. However, the mere presence of this fruit in no way means that a developed fruit culture existed. While it is conceivable that people may have planted saplings near their settlements, they were certainly not aware of grafting and pruning techniques. But the thick layers of pressed apple peels discovered by archaeologists indicate that the ancient Germanic tribes knew how to ferment fruit juice.

The only fruit-related finds from pre-Roman Britain are wild raspberries (near Stonehenge and in Shippea Hill, Cambridgeshire), sour cherries from the Neolithic and Middle Bronze Age (at Notgrove Long Barrow, Gloucestershire, and in Haugh Head, Northumberland), and crabapple pips, again from the Neolithic period, in Windmill Hill in Wiltshire. All these fruits were harvested, but not cultivated.

In any case, the British climate didn't really lend itself to the early cultivation of fruit. Tacitus writes:

The climate is wretched, with its frequent rains and mists, but there is no extreme cold. Their day is longer than in our part of the world... The soil will produce good crops, except olives, vines and other plants which usually grow in warmer lands. They are slow to ripen, though they shoot up quickly, both facts being due to the same cause, the extreme moistness of the soil and atmosphere.

left
Bacchus was the Roman god of wine, fruit, and agriculture, equivalent to Dionysus in ancient Greek culture, sixteenth century.

Despite the gray, damp weather they encountered, the Romans were ready to bring their experience from southern lands to Britain's rather dreary shores. While Tacitus claimed that grapevines would not thrive in the British Isles, archaeologists have found traces of Roman vineyards in seven locations there, including four in Northamptonshire. The Nene valley in this region was apparently a wine-making hotspot: near the village of Wollaston, excavations turned up ancient vineyards that extended for at least thirty acres (twelve hectares) and utilized the same Mediterranean methods that Pliny the Elder and Columella described.

The bedding trenches found at one site were extensive enough to contain four thousand vines that in turn would have produced up to 2,650 American gallons (10,000 liters) of wine each year. Most of these were probably sweet, fruity, brown-colored beverages made from grapes that did not have the chance to ripen completely, and the pressed juice would have been sweetened with honey. The resulting mixture was fermented in amphorae or barrels and had to be drunk within six months. Since the grapes were harvested in late September, this wine would have been available only in the winter and spring months. The rest of the year, wine enthusiasts had to make do with alternatives prepared from raisins.

Of course, Romans also brought other types of fruit from Italy and other parts of the Continent to Britain. Discoveries in West Sussex testify to the transfer of Roman gardening culture to this corner of the Empire. At Fishbourne, archaeologists found the remains of one of the largest Roman villas north of the Alps. Its main attractions are impressive floor mosaics, but the gardens have been partially reconstructed as well. While we can't know for sure whether they contained fruit trees, it seems likely that the gardeners at least experimented with them. After all, fruit cultivation was highly advanced, and the climate in southern England at

that time was especially mild—although still challenging for species originating in the sunny south.

While Tacitus does not explicitly name fruit such as apples and pears, the Romans almost certainly planted such trees in Britain. Kent is famous for its orchards, and the first of them probably originated during the Roman occupation. Mulberry seeds found in a dig in Silchester must also have come from imported plants, since mulberry bushes were not native to the British Isles. Sweet chestnuts and walnuts came with the Romans as well. In short, the Romans left their imprint on Britain's natural world, just as they did throughout their empire: the plant population changed, and from that point on it followed a new path of development. Part of this development would include continued exchange among the lands of Europe under entirely new conditions—those created by a network of monasteries and convents.

above
Roman mosaic discovered at Hinton St. Mary, Dorset, depicting a Christ-like figure flanked by two pomegranates.

— 6 —

Earthly Paradises

URING THE SIXTH THROUGH the fifteenth centuries, Catholic orders—especially the Benedictines—planted legendary gardens in central Europe and on the British Isles. For these self-supporting religious communities, a garden was a vital source of fruit, vegetables, and herbs for both seasonings and medicinal purposes.

Just how important these plant products were to the monks and nuns becomes even clearer when we remember that most religious orders, especially the early ones, followed a largely vegetarian diet. Milk, eggs, and cheese were allowed outside of Lent. During Lent, fish was permitted, and many monasteries and convents maintained fishponds, often stocked with carp. Meat from four-legged animals, in turn, was avoided. Cookbooks of the period, such as the famous *Liber de arte coquinaria* (*The Art of Cooking*), published by the Italian master chef Maestro Martino in 1460, reveal that fresh fruit was rarely served. Instead, it was far more typical to cook or boil the fruit. But of course we can't rule out the occasional snack snuck straight from the tree.

In the biblical paradise, the seasons never changed and no one needed to perform laborious gardening tasks. Everything grew luxuriantly on its own. Real gardens, of course, were quite different. The romantic vision of nuns and monks working within protective stone walls in silent harmony with nature may have held true during the early medieval period, when the gardens were still relatively small and manageable. Later, however, monasteries and convents grew into commercial enterprises that produced goods not only for their own use but for high-paying external customers. Hiring additional gardeners became essential. Pforta, a twelfth-century Cistercian abbey in what is now the eastern part of Germany, provides an example: records there document that a master gardener was in charge of the orchard. Many monks and nuns were aristocrats and joining a religious community did not necessarily eliminate the privileges they enjoyed. Furthermore, in line with the Benedictine maxim *ora et labora* ("pray and work"), much of the typical day for order members was filled with praying and reading. All this meant that hired hands often performed the demanding physical labor, even though the order was always in charge.

One excellent source of information on the form that medieval gardens took is the plan of a Benedictine monastery found by scholars in the abbey library of St. Gall in Switzerland. Probably drawn in 816, it was rediscovered in 1604 but not examined more closely until the middle of the nineteenth century.

The circumstances surrounding its creation are not clear and the document may very well be a draft that was never realized. Nevertheless, the drawing provides valuable clues about how such a monastery was set up. In addition to showing diverse buildings, it

contains labels indicating the locations of a *herbularius* or garden with medicinal herbs, a *hortus* or vegetable garden, and a *pomarius* or orchard. One detail about this last feature seems strange from our point of view: it was also the cemetery. But this double duty was certainly no coincidence. A fruit tree's annual cycle of winter dormancy, blossoming, and fruit formation served as both an analogy of human life and a symbol of resurrection. A popular legend in the Middle Ages emphasizes this connection: it relates how a scion was grafted to a dead tree trunk and brought it back to life.

The plan from St. Gall shows the graves of the monks in rows arranged symmetrically around a high cross in the middle. The Latin inscription reads: "Around the cross lie the brothers' lifeless bodies; when it glows in the light of eternity they will rise up to life." On the cross itself, the symbol of deliverance, stand the words: "Of all Earth's trees the cross is the most sacred, as on it the fruit of eternal salvation spreads its sweet fragrance."

The trees are all symbolized with the same precisely drawn branching ornament, but a label indicates their specific species: *mal* (apple), *perarius* (pear), *prunarius* (plum), *mispolarius* (medlar), *castenarius* (chestnut), *persicus* (peach), *avellenarius* (hazelnut), *nugarius* (walnut), *murarius* (mulberry), *guduniarius* (quince), *amendelarius* (almond), *laurus* (laurel), *sorbarius* (rowan), *ficus* (fig), and *pinus* (pine). Apples, pears, and quince appeared most frequently.

Fortunately for us, the writings of Walafrid (808–849), a Benedictine monk, botanist, and author from the Carolingian period, have survived to provide additional information. Walafrid—also known as "Strabo," the squinter, apparently due to his crossed eyes—joined the monastery on the island of Reichenau in Lake Constance as a boy and later served there for a decade as the abbot. In *On the Cultivation of Gardens*, a poem about gardening, he addresses his teacher, Father Grimaldus:

> You sit in the fenced enclosure of your small garden, in the shade of your fruit trees with their leafy crowns, where peaches rest at the top in the broken shadows, and where the boys, your cheerful pupils, pluck the pale, tender, fuzzy fruits and place them into your palms, as they try to encircle the large peaches in their own hands.

What specific types of fruit were grown in a garden like this? Documents like the *Capitulare de Villis*, a set of agricultural laws under Charlemagne, provide a few clues. Charlemagne's agricultural laws aimed to ensure a good supply of food across his empire, including a balanced apple harvest, with a mix of sweet and tart apples as well as apples meant to be eaten quickly and others that could be stored for longer periods. None of the apple varieties listed in the document still exists. They were simply way stations in a long history of cultivation, and as new forms arose they vanished. All that remains are hints as to what those apples might have been like. The Grey French Rennet, for example, is considered an early form of the Rennet or Reinette apples still grown today. It can be traced back to the Cistercian monastery of Morimond in northern France, which was founded in the twelfth century.

As we have seen, the plan of St. Gall features fig trees, which are traditionally found in the Mediterranean region. Is it really likely that plants from southern climes grew in the gardens of central European monasteries? Although the so-called Medieval Warm Period only lasted from 900 to 1300, fifteenth-century sources reported that the Franciscan monks in the monastery at Dresden had fig trees in their garden. Supposedly Duke Albert of Saxony brought them back from his 1476 pilgrimage to the Holy Land. Some types of citrus fruit were known during the Middle Ages and may have found a place in monastery gardens as well. The Dominican monk and university-educated scholar Albertus Magnus (ca. 1200–1280) wrote a book about fruit that mentions the blossoms and scent of the citron (*Citrus medica* var. *Cedro*) and bitter orange (*Citrus aurantium*).

Medieval accounts indicate that cloisters were sometimes surrounded by a ring of fruit trees. The Cistercians in particular considered it important to claim and cultivate additional land beyond the confines of their walls. They were especially skilled in this area and viewed such efforts as part and parcel of their work.

The Bible commissioned by King Wenceslaus IV at the end of the fourteenth century in Prague contains a number of illustrations—including one that gives us an idea of what such an orchard outside a monastery might have looked like. A circular fence made of woven brushwood is interrupted by a small roofed entryway. Behind it are about a dozen trees and bushes, some heavily laden with fruit. Nine people are busily working there: some are picking and collecting the fruit, while another appears to hug one of the trees—presumably he's shaking it. The oldest image of a cloister garden on the British Isles was drawn in 1165 as part of the plan

of Christ Church in Canterbury. The "pomerium" shown in the plan consists of two gardens with trees that cannot be clearly identified but are definitely bearing fruit.

The order of the Carthusians deserves special mention. At their monasteries it was customary for each monk to have his own garden plot, which could be up to one thousand square feet (one hundred square meters) in size. Wood or stone gutters ensured a supply of water, and the decision of what to plant was up to the monk in charge. There was just one rule: trees could not grow so tall that they blocked the sunlight from the neighboring plots.

In addition to trees, the plan of St. Gall shows beehives, which were likely a part of these orchards from early on. After all, bees not only helped to pollinate the plants, but also made wax for candles and honey for eating and cooking. It is hard to think of a more symbiotic relationship than the one between orchards and bees. There was an element of symbolism for the monks, too. As bees were never observed mating, they were associated with celibacy, which made beeswax the perfect substance for the candles used to light the monks' places of worship.

What was the atmosphere of these cloister gardens like? The development of religious orders was heavily influenced by the practices of early Christian hermits, whose solitude made a life of contemplation possible. In the quiet and relative isolation of a cloister garden, blossoms and scents are especially powerful. *Vitae Patrum: The Lives of the Fathers,* an account by the Frankish chronicler Gregory of Tours, written in about 580, describes an abbey in Clermont founded by St. Martius, a lifelong servant of God who lived to the grand old age of ninety:

> There was a garden for the monks, filled with a variety of vege-
> tables and fruit trees; it was pleasant to look at and happily
> fertile. The blessed old man often used to sit in the shade of its
> trees while their leaves whispered in the wind.

There and elsewhere, the brothers or sisters read the works of ancient authors like Vergil and Ovid, interpreting them in line with their own Christian views. Plato, Epicurus, and Theophrastus had met with their students in gardens to philosophize together, and this tradition continued in the Middle Ages. Reading antique descriptions of gardens and nature influenced how the monks and nuns saw the plant life around them.

An account from the fifteenth century shows yet another way that orchards could benefit a religious community—in this case, the monks of Clairvaux. Long before the dawn of modern medicine, brothers who were ailing could receive a range of comforts:

left
An enclosed orchard from the Bible commissioned by King Wenceslaus IV of Bohemia, late fourteenth century.

*Many fruit-bearing trees of various species… by their near-
ness to the infirmary afford no small solace to the brothers in
their sickness: a spacious promenade for those able to walk, an
easeful resting-place for the feverish… The sick man sits on the
green turf… While he feeds his gaze on the pleasing green of
the grass and trees, fruits, to further his delight, hang swelling
before his eyes… Thus for a single illness God in his goodness
provides many a soothing balm: the sky smiles serene and clear,
the earth quivers with life, and the sick man drinks in, with eyes,
ears and nostrils, the delights of colour, song and scent.*

The role of cloisters in advancing plant cultivation cannot be
underestimated. The fact that they shared information in a lively
network led to better fruit varieties, and plants from more southern regions were systematically introduced and disseminated.
Monks and nuns were also the first people to cultivate strawberries and raspberries instead of simply collecting them in the forest.

In recent years, French archaeologists made the exciting
announcement that they were attempting to reconstruct the papal
gardens of Avignon, which existed nearly seven hundred years ago.
Seven popes held court in the city of Avignon between 1309 and
1376, a period when the papacy was under the sway of the powerful French kings and Italy was engulfed in conflict.

The first papal garden in Avignon was created in 1335 under
Pope Benedict XII. Laid out along the eastern side of the papal palace from the late Gothic period, it covered about 21,500 square
feet (2,000 square meters) and was surrounded by thick walls on
all sides. A team of researchers under archaeologist Anne Allimant-
Verdillon discovered seeds and plant remains from both vegetable and fruit species, including grapevines and orange trees that
were grown in ceramic pots. The popes that followed each made

changes to suit their own taste, adding trellises for vines, decorative trees, and even arbors and benches. It's easy to imagine people moving through and enjoying the resulting spaces.

Gardeners were responsible for caring for these grounds, with assistants—both male and female—to perform simple tasks like weeding or collecting snails and other pests. In fact, the gardens were the only part of the palace where women were allowed to work. Pope Gregory XI was persuaded to return to Rome in 1376, and the gardens soon fell into neglect. The team working today to bring them back to life hopes that some of the seeds recovered from the layers of fourteenth-century earth will sprout.

above
Beehives as shown in the *Tacuinum Sanitatis (Maintenance of Health),* late fourteenth century.

This leads us to a related question: Is it possible to recreate a cloister garden exactly as it once was? The idea of starting from scratch with the plan of St. Gall hardly seems feasible. But nearly thirty years ago, an attempt began that was almost as ambitious—and that, fortunately for us, continues today.

In 1991 Sonia Lesot and Patrice Taravella, both architects from Paris, came upon the abandoned priory Notre-Dame d'Orsan, a relatively modest place that had once belonged to the Abbey of Fontevrault in the Loire valley. Notre-Dame d'Orsan was founded in 1107 by a man named Robert d'Arbrissel. Its severe-looking twelfth-century buildings with high red-tile roofs surround a courtyard on three sides. When Lesot and Taravella discovered them, they were run down, and the pigstyes, chicken house, and other outbuildings were in ruins. Undeterred, the architects bought not only the priory but fifty acres (twenty hectares) of surrounding forest and meadows. They spent the first years just clearing out the rubble, getting organized, and preparing for the next steps. And they were not so naive as to believe that they could recreate the complex exactly as it had been. After all, they didn't even know what it had originally looked like.

Medieval miniatures, tapestries, and texts provided inspiration for their work. And they made a few fundamental decisions. Wood would be the preferred building material whenever possible. The garden work would be performed by hand using simple tools. Traditional substances would serve as fertilizer: manure, pulverized horn, and dried blood. They chose a combination of copper sulfate and quicklime—known as Bordeaux mixture—as a fungicide even though, strictly speaking, it was invented in the nineteenth century. They also decided to use a modern watering system so that they could devote more time to the plants themselves.

right
An illustration from a fifteenth-century edition of the French medieval poem *The Romance of the Rose* showing a pleasure ground of orchards.

Like an actual medieval cloister garden, the garden at Notre-Dame d'Orsan contains a wide assortment of plants, but fruit trees are well represented. The kitchen garden features seventy plum trees designed into a hedge labyrinth symbolizing the road to Jerusalem. To ensure a supply of ripe plums at different times throughout the late summer, Lesot and Taravella chose four different varieties: Sainte-Catherine, Mirabelle de Nancy, Reine-Claude dorée, and Quetsche d'Alsace. The walls of the outbuildings are covered with apple and pear trees trained into U shapes. Their names—including Transparente de Croncels, Reinette du Mans, and Calville blanche—are evocative but they are all cultivated varieties that did not exist eight hundred years ago.

A small vineyard is present as well, with Chenin blanc grapes growing on vines that climb up hurdles: fence panels woven from pliable young chestnut whips. In some years, the harvest can be turned into Chenin blanc wine, with its distinctive honey aroma.

A pasture to the northeast of the property where three pear trees still stood has become the main orchard. Trees bearing twenty-three different types of apples are laid out there in an arrangement based on a pattern known as the quincunx. Anyone who has ever played a game using dice has seen a quincunx: it's a square with a dot at each corner and another in the middle—in other words, the symbol for five. The first-century Roman rhetorician Quintilian describes this pattern in his seminal work *Institutes of Oratory*, which was published around 95 CE:

> What is more beautiful than the quincunx, that, from whatever direction you regard it, presents straight lines?... Shall not beauty, then, it may be asked, be regarded in the planting of fruit trees? Undoubtedly; I should arrange my trees in a certain order, and observe regular intervals between them.

He also believed that planting trees in a regular pattern supported their growth and health, "as each of them then attracts an equal portion of the juices of the soil."

Berry bushes line the road separating the cloister from the orchard. In the herb garden two olive trees provide shade in the summer; in the winter they are covered against the cold. Careful observers can even spot small orange trees.

The abbey has risen from the ruins. Thanks to the efforts of Sonia Lesot and Patrice Taravella—and the head gardener, Gilles Guillot—we can imagine how the monks who lived here might have experienced this place. Notre-Dame d'Orsan is a monument to a time when God and nature guided human activities. And the garden holds a secret as well. Before his death, the cloister's founder arranged for his body to be entombed in a cathedral, but his heart was to stay in Orsan. Where else could it have been buried?

The Cloisters, as the name suggests, is another interesting reconstruction of a medieval cloister. Located on the north end of Manhattan, it stands on a small hill in a park above the Hudson River. The structural elements of the complex, which opened in 1938, were brought together from various sites in France. The associated Bonnefont Cloister garden, surrounded by colonnades that echo the peristyle gardens of Roman times, contains grapevines, espaliered pear trees, and a few quinces. Like other "medieval" gardens that exist today, it is just an approximation of its historical models, but it can still evoke the atmosphere of these gardens and give us some sense of what it must have been like to spend time under the fruitful trees that filled them.

THE FOUR SEASONS OF THE HOUSE OF CERRUTI, a Veronese codex richly illustrated with miniatures, is a treasure trove of insights about how people understood and cultivated a wide range of plants and trees in medieval times. This remarkable document, also known as *Tacuinum Sanitatis* or *Maintenance of Health*, defies present-day categorization; it is based on an all-encompassing concept of health and diet that seems very foreign to us. The knowledge stored in its pages originated far earlier than the book itself: it developed from the Arab medical tradition which, in turn, drew on the classical medicine of antiquity as practiced by legendary figures such as the Greek physicians Dioscurides (40–ca. 90) and Galen (129–ca. 216).The ideas of Ibn Butlan, a Christian physician born in Baghdad in the early eleventh century, also had an influence.

The book's 206 illustrations are believed to be the work of Giovannino de' Grassi, a Milanese artist. They depict the mores and customs of northern Italy in the late fourteenth century. Specific trees and plants play important roles, along with animals, spices, types of water, and even winds. Each picture conveys a recommendation, and although today we may doubt their effectiveness, they certainly provide a valuable view of how people at the time regarded nature.

While a complete orchard never appears, the illustrations provide separate glimpses that help us fill in the full picture. They are organized by season. The entry on the pear—*pirna*—provides a sense of the approach:

> *Fragrant and perfectly ripened pears generate cold blood and are therefore suited to those with hot temperaments, in the summer, and in southern regions. They are salubrious for people with weak stomachs, but harmful to the production of bile. The harm can be remedied by chewing cloves of garlic after the meal.*

It goes on to claim that according to Dioscurides the ashes of a pear tree dissolved in a drink will neutralize poisonous mushrooms. In fact, the reader is told, "by cooking wild pears together with mushrooms, you will avoid any harmful effects." Let's hope that credulous readers didn't put this theory to the test.

right
With *The Golden Age*, ca. 1530, the German Renaissance painter Lucas Cranach the Elder created a vision of an ideal world where all people had to do was relax, enjoy themselves, and eat plentiful fruits.

CLOISTERS AND THEIR GARDENS might have been attempts to create a world governed by biblical principles, but they were just pale
copies of the original. For centuries, the Garden of Eden retained
its power to fascinate. It developed an independent existence in
the collective imagination and eventually, after a series of detours,
left its mark in reality. In his *Decameron* (1348), the Italian writer
Giovanni Boccaccio (1313–1375) described a garden of flowers and
fruit trees that, in its loveliness, reached near-Edenic proportions:

> *It was sprinkled all over with perhaps a thousand kinds of
> flowers and surrounded by luxuriantly growing, bright green
> lemon and orange trees that were covered with blossoms as well
> as both mature and ripening fruit, and that provided a pleasant
> shade for the eyes as well as delightful odors.*

Yet Boccaccio is still clearly describing a garden that could
actually exist. The Jesuit scholar Athanasius Kircher (1602–1690)
goes quite a step further. Not content to simply consider the idea
of Eden, he attempted to capture it in a much more concrete form:
a map. This cartographic drawing shows the garden's layout as
seen at an angle from above and locates it in Mesopotamia. The
garden is laid out in the shape of a large rectangle surrounded by a
wall and filled with trees. What appears to be a spring in the middle feeds four rivers that branch out in different directions (and
continue flowing beyond the garden wall). Adam and Eve are vis-
ible under a particularly large tree.

Kircher didn't choose this form for the garden out of the blue.
It represents the archetype of a quadratic garden divided into
four sections: a design with roots in pre-Biblical times found in
places like the gardens of the Persian ruler Cyrus II in his capital
of Pasargadae. Later this basic geometric structure lived on in

the enclosed gardens of medieval churches and cloisters. Islamic gardens were no less indebted to paradise as a model, and examples from Andalusia to India echo this form as well. While these quadratic gardens did not invariably contain fruit trees and were certainly not the only place to find them, they established an important pattern that persisted through millennia.

Kircher was no eccentric. Seventeenth-century Christians took the Bible's account of the world's origins literally to an extent that we, in our more scientific age, can hardly imagine. And in the British Isles an entire series of authors essentially created a movement to apply this biblical story to actual orchard design. One example was the Puritan John Milton (1608–1674), who had a very precise idea of paradise—he considered it to be a large area in the East, surrounded by the "goodliest trees laden with fairest fruit." What's more, he imagined specific tasks associated with this realm: for the trees to be bountiful someone had to "lop," "prune," "prop," or "bind" them. But Adam and Eve's labors were well rewarded because the fruit was "burnished with golden rind" and "of delicious taste."

Even earlier, in the sixteenth century, various writers were waxing eloquent on the connections between fruit—especially exotic species—and the original Garden of Eden. A number of books that appeared during this time addressed the question of what arrangement of plants and fruit trees would make a garden genuinely paradisiacal.

In his first book, *Paradisi in Sole/Paradisus Terrestris*, the royal physician John Parkinson (1567–1650) provides detailed descriptions of the plants and trees for a flower garden, a kitchen garden, and an orchard. (The book's title, which translates as "Park-in-Sun's Terrestrial Paradise," is a pun on his name.) Parkinson's Eden thus consisted of three gardens where everything found on earth grows.

One of the plates, "Modell of an Orchard," shows a geometric lay-out of trees that could be repeated or continued in every direction. In the resulting pattern, the trees create small avenues and the branches form arches connecting them.

Ralph Austen (1612–1676), from Leek, Staffordshire, operated a commercial nursery and spent the second half of his life in Oxford, where he established an apple wine business in 1659. He was con-vinced that new crop plants and gardening methods could help solve social problems like poverty and unemployment. And he found a unique way to express the "dignity and value" of fruit trees, investing them with meaning beyond their immediate usefulness. In *The Spiritual Use of an Orchard or Garden of Fruit Trees* he wrote:

> *The world is a great library, and fruit trees are some of the books wherein we may read and see plainly the attributes of God, his power, wisdom, goodnesse &c. and be instructed and taught our duty towards him in many things, even from fruit trees: for as trees (in a metaphoricall sense) are books, so likewise in the same sence they have a voyce, and speak plainly to us, and teach us many good lessons.*

left
Medlars, a once-popular fruit, are uncommon today, sixteenth century.

Strictly speaking, this publication was a pamphlet containing part of his 1653 work *A Treatise of Fruit-Trees*. The title page fea-tured a reference to Song of Solomon 4:12–13: "A garden inclosed is my sister, my spouse; a spring shut up, a fountain sealed. Thy plants are an orchard of pomegranates, with pleasant fruits." The message was that profits and pleasures should go hand in hand—an idea driven home quite literally by the presence of both terms on the cover, joined by a drawing of a handshake. Everything in perfect harmony: What could be better?

For Austen, a productive orchard was the key to paradise. Echoing him is Stephen Switzer, the author of *The Practical Fruit-Gardener* (1724), who was trained in London's Brompton Park nursery, the leading such business at the time. Switzer wrote:

> *A well-contriv'd Fruit-Garden is an Epitome of Paradise itself, where the Mind of Man is in its highest Raptures, and where the Souls of the Virtuous enjoy the utmost Pleasures they are susceptible of in this sublunary State.*

He combined this credo with precise directions for designing what he considered to be the perfect garden.

William Lawson's earlier *A New Orchard and Garden* (1618) was an extremely popular book about gardening written in a style that, to critics in his day, recalled the language of the King James Bible. Here's a sample:

> *Of all other delights on earth, they that are taken by Orchards are most excellent, and most agreeing with nature…What can your eye desire to see, your ears to hear, your mouth to taste, or your nose to smell, that is not to be had in an Orchard, with abundance and variety?*

For Lawson, an orchard was a landscape taken straight from paradise:

> *And in mine own opinion, I could highly commend your Orchard, if either through it, or hard by it, there should run a pleasant River with silver streams; you might sit in your Mount, and angle a speckled Trout, sleighty Eel, or some other dainty Fish. Or Moats, whereon you may row with a Boat, and fish with Nets.*

Many Elizabethan or Stuart gardens really did contain an artificial hill or "mount." Stairs or winding small paths led to its highest point, which offered a pleasant view. We should not underestimate the importance of the opportunities such gardens provided to get some exercise. One astounding passage makes this even more explicit:

> *To have occasion to exercise within your Orchard, it shall be*
> *a pleasure to have a Bowling-Alley, or rather (which is more*
> *manly, and more healthful), a pair of Buts, to stretch your Arms.*

The idea of designing orchards in precise patterns as such a clear reflection of Christian spiritual beliefs was a short-lived phenomenon among a relatively limited group of people. The gardens planted later continued to follow the same rational principles but were not intended to emulate a heavenly plan. But the idea that gardens were places especially suited to promoting salvation did not simply disappear from one day to the next. Instead, it lived on from age to age, taking new forms shaped by the beliefs of the time.

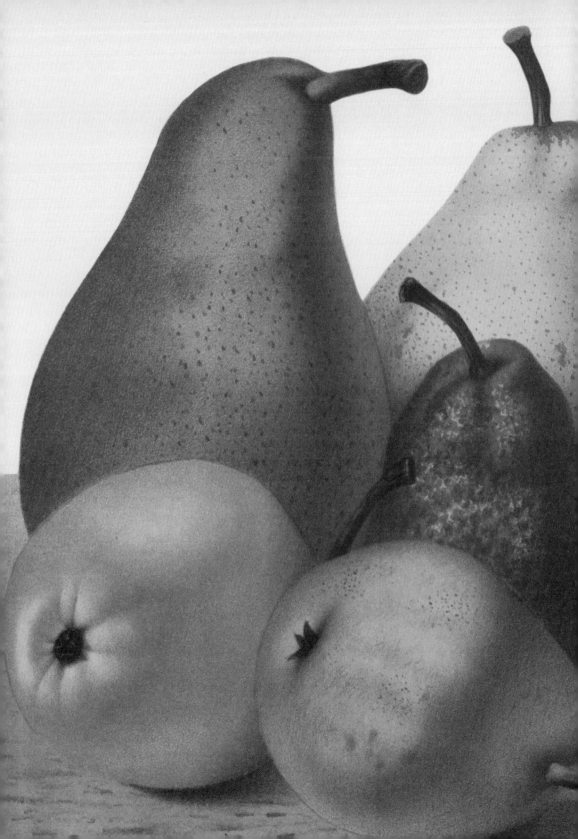

— 7 —

Pears for the Sun King

T HE *POTAGER DU ROI* at Versailles translates as the "king's kitchen garden," but the name, conjuring up beds of vegetables and herbs as it does, is quite an understatement. The palace grounds were designed by André Le Nôtre, the head gardener of Louis XIV, the Sun King. In 1678, Jean-Baptiste de La Quintinie—a lawyer by training but a gardener at heart—began to lay out an expansive twenty-two-acre (nine-hectare) garden south of the palace to provide the court with fruit and vegetables. The earlier such garden was too small for the task. The swampy stretch of land chosen for this project did not seem promising: sources from the time describe it as a stinking bog. It took several regiments of Swiss guards to drain it and fill it with fertile soil.

Versailles, near Paris, became the residence of the French court and the seat of the government in 1682. The new kitchen garden was finished the following year. Like the rest of the palace, the kitchen garden was designed to serve the king's public image. The entire ensemble functioned as a sophisticated work of art that expressed the desire—typical for the age—of demonstrating dominance over nature. Louis made La Quintinie the director of his *potager* and often had La Quintinie escort him through the garden. The side entrance he used was fit for an absolute monarch: it was a specially built gate richly decorated with cast-iron ornaments. When important guests such as the doge of Venice or the ambassador of Siam were visiting, Louis made sure they toured the garden, too.

Part of the surrounding wall was expanded to provide a terrace where the king could walk and look down at the bounty below. One interesting architectural feature was the inclusion of twenty-nine enclosed spaces on the exterior wall that faced south, east, or west. Because these parcels were shielded from the wind, they developed a relatively warm microclimate on sunny days even in the cooler parts of the year. As a result, they were the perfect places for cultivating fig, peach, and apricot trees, which are sensitive to the cold. Melons, strawberries, and raspberries (of the type Sucrée de Metz) also supposedly grew there. The branches of the trees were not simply allowed to grow as they pleased; instead, they were trained into espaliers—some of them quite fantastical—that harmonized with the highly artificial appearance of the entire palace and its grounds. Of course, La Quintinie also knew that forcing trees into a fan shape meant that more sunlight would reach the branches.

La Quintinie must have been under enormous pressure to please his highly demanding client and supply a court of several thousand people with the desired quantity of beautiful and choice

previous page
Pears come in many flavors, colors, and shapes, an illustration from 1874.

fruit. In summer, the garden was expected to provide four thousand figs each day. Among La Quintinie's other remarkable feats
was cultivating plants outside of the season when they normally
grew. Legend has it that he was able to delight the courtly palates
with strawberries in January. Even if the story is true, the berries
were certainly not particularly sweet. At the time the garden was
founded, greenhouses hadn't been invented yet; they arrived on
the scene fifty years later. But La Quintinie had other tricks up his
sleeve. He placed glass domes over plants that grew early in the
season and had fresh horse manure spread over the earth to warm
it. Other rulers were soon interested in his prodigious talents, and
he is said to have turned down a competing offer from the king
of England.

During his time in Versailles, La Quintinie even found the time
to write a handbook on orchards and vegetable gardens, published
in English as *The Complete Gard'ner; Or, Directions for Cultivating
and Right Ordering of Fruit-Gardens and Kitchen-Gardens*. It is no
accident that in addition to covering many aspects of botany, this
remarkable book describes five hundred types of pears. La Quintinie wrote: "It must be confessed that, among all fruits in this
place, nature does not show anything so beautiful nor so noble
as this pear. It is pear that makes the greatest honor on the table."

The Sun King had a special preference for this fruit, and especially for the fragrant, sugar-sweet Bon-chrétien d'hiver that had
long been considered a symbol of the rising sun. The mythos surrounding this fruit, with its buttery consistency and musky aroma,
took on a life of its own. When an orchard operator named Williams
presented the pear at a fruit exhibition of the Horticultural Society of London in 1816, it was dubbed Williams Bon Chrétien, and
from there it conquered the world. It had already found its way to

Boston, and after Enoch Bartlett acquired the land where it grew, the pear was given his name. Bartlett remains the most common name for this pear today in the United States and Canada.

During the following centuries, the garden of Versailles passed through several different phases. One especially memorable innovation came in 1735. La Quintinie's successor, Louis Le Normand, was able to proudly present the first pineapples cultivated in heated greenhouses to Louis xv. Even banana trees were planted there. In The Lost Gardens of Heligan, part of an estate rediscovered a few decades ago in Cornwall, England, and then reconstructed, pineapples are still cultivated in special pits kept warm with horse manure.

The seventeenth and eighteenth centuries in France were a golden age for orchards, and the country set the pace for developments in other parts of the world. By the end of the seventeenth century, an extensive network of tree specialists and fruit farmers stretched across large parts of Europe. French gardeners—mostly from the region around Paris, the center of fruit production at the time—went to the British Isles and Germany, and their British counterparts came to France. The sales and exchange among them were lively, and hot pursuit of the most unusual and attractive varieties was the norm.

It didn't take long for this network to expand to North America. In the eighteenth century, the plum known as *Zwetschge* in Germany and *prunier de Damas* in France grew on Montréal's fruit farms alongside local varieties, as did Reinette and Calville apple trees. A nursery in France owned by the Andrieux family boasted of its contacts around the world. However, the "monstrueuse Reinette du Canada" that Andrieux offered to customers as a grafted tree most likely originated in Normandy. The name was probably just a marketing trick to make the plant seem more interesting and exotic.

Pears, peaches, and figs were particularly celebrated during this period. People claimed to be bewitched by the scent. "Does the musk or amber exist that surpasses the aroma of a good pear or a good peach when they are fully ripe?" asked René Dahuron in his 1696 book *Nouveau traité de la taille des arbres fruitiers* (*New Treatise on Pruning Fruit Trees*). For Dahuron, a student of La Quintinie, the question was purely rhetorical. The plum, in contrast, had few fans; "he'd rather have two eggs than a plum" was a common saying in France at the time. Space for espaliers growing on walls was considered too valuable to waste on plums.

Why was this fruit so unpopular? Its primary "failing" was the fact that it lacked a reputation as uncommon or exotic. The plum was also notorious for its laxative qualities. And the mature trees were considered ugly—they were seen as sparse and misshapen. Of course, this judgment was unfair, especially since the plants were not trained as espaliers and of course did not exhibit the regular, elegant form of the highly cultivated pear and peach trees. What's more, plum trees were easy to care for and well adapted to the climate in central Europe. Raising them required no special talent—in fact, the trees often grew wild. But despite their lowliness, artists appreciated plums for their range of colors. In glowing shades of yellow, red, green, purple, and black, they enliven a variety of still lifes.

left
Bon Chrétien pear, 1853.

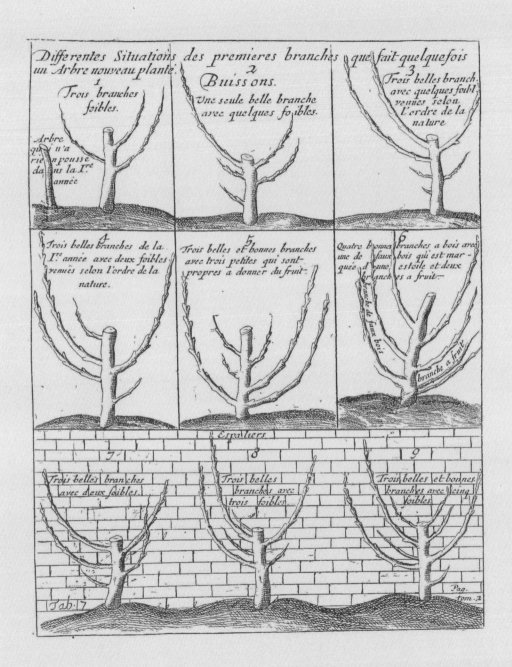

The competition between apples and pears, in turn, was especially keen. This is surprising in some respects, because pears are
much more problematic: they must be picked before they are ripe,
and because they are more delicate they are harder to transport.
But the clear preference for pears shared by the Sun King and his
head gardener placed apples at a disadvantage. It was quite a fall
for this fruit, since a century earlier writers were still singing its
praises. One such paean can be found in *L'agriculture et maison
rustique* (1564), where Charles Estienne and his son-in-law Jean
Liebault wrote, "The apple tree is the most necessary and valuable
of all; for this reason Homer was already calling it the tree with the
beautiful fruit in his day." And the apple continued to have fans
even when the Versailles gardens were setting the tone.

LA QUINTINIE REMAINED FRANCE'S top authority on fruit farming
until the end of Louis xv's reign. Many other authors simply copied
what he had written, and some didn't even bother trying to cover
their plagiarism tracks. Genuinely new books demonstrating a more
advanced understanding of plant processes, like *Traité des arbres
fruitiers* (*Treatise on Fruit Trees*) by Henri-Louis Duhamel de Monceau,
did not emerge until the second half of the eighteenth century.

While applying the latest insights to grafting and cultivation
was already quite an accomplishment, the masters of fruit farming also had another important skill. They were able to organize
the varieties in their orchards so that they would ripen at different times, ensuring a continuous supply of fruit. Of course, this
feat was not possible for all fruit species. However, some sources
claim that as early as the second half of the seventeenth century,
France's leading gardeners were able to provide pears nearly year-
round. The early-ripening D'Epargne and De Jargonelle opened the
season, maturing in the valley of Montmorency from mid-June to

the beginning of July. They were followed by À Deux Têtes, Royale, and the large summer cooking pear, De Vallée. The fall brought the lovely, luscious varieties Messire-Jean, De Tantbonne, D'Angleterre, and De Bergamote. Meanwhile, the winter pears Poire d'hiver, Vigoureuse, De Martin-sec, and De Dangobert lay waiting in the storage shed or fruit cellar, slowly reaching their optimal ripeness. When everything went according to plan and the weather behaved as expected, fresh pears could be sold and eaten from the middle of June until the following May—that's an astonishing eleven out of twelve months. Of course, the price varied over the course of the year, and we don't know how the last pears in May tasted. It seems doubtful that they would have met today's high standards for flavor and texture.

Orchards were living laboratories: orchardists reinvigorated existing varieties and acclimated new species, all while managing their sometimes-challenging relationships with their patrons and masters. Often fruit trees had to share the available space with grapevines, vegetables, and plants used as animal feed. Other trees and flowers might be present as well. Such diversity in a limited area created a favorable microclimate in which watering and fertilizing the plants was relatively simple.

The practice of planting fruit trees and other plants in organized patterns was not only aesthetically pleasing but made good use of space, and careful pruning opened the interior of the trees to the light the fruit needed to ripen and allowed sunlight to reach the plants below. But sometimes unscrupulous gardeners took advantage of the need for pruning to cut down more branches than necessary—and reap ill-gotten gains in the form of valuable firewood. One orchard owner accused his gardener, Pierre Bonnel, of just such shenanigans around 1700. Bonnel's defense was that

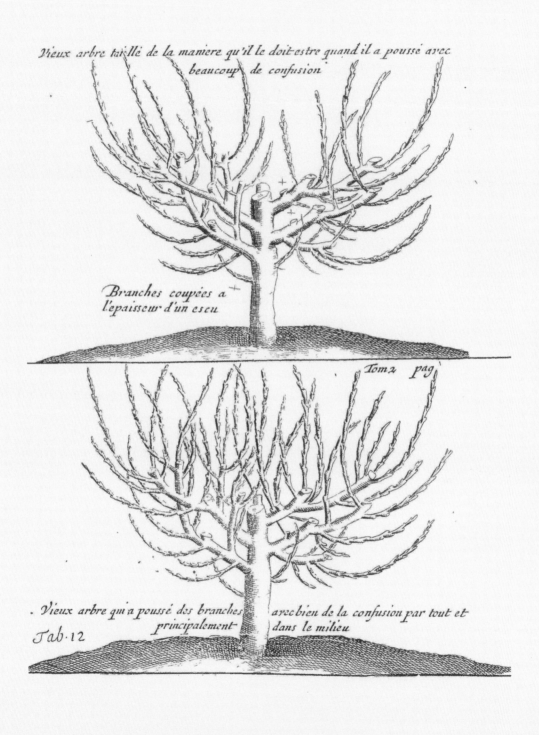

Vieux arbre taillé de la maniere qu'il le doit-estre quand il a poussé avec
beaucoup de confusion

Branches coupées a
l'epaisseur d'un escu

Tom.2 pag.

Vieux arbre qui a poussé des branches avec bien de la confusion par tout et
principalement dans le milieu

Tab·12

"eight hours before Christmas" he had sawed off the branches of some of the apple trees to ensure the grapevines growing under them would get enough air. Perhaps this was true, or perhaps Bonnel was simply looking to add a little extra cheer to his holiday.

The risk of "theft" and harm existed from another side as well. In central Europe and beyond, both orchards and vineyards were usually surrounded by fences, hedges, and stone walls to protect them from the depredations of domestic animals. When fruit trees grew in more open terrain, unprotected from grazing livestock, their trunks were surrounded by sticks, thorny brushwood, or strips of old cloth that fluttered in the breeze—whatever was considered necessary to keep the foe at bay. In some places animals were allowed to graze in orchards as long as there was no conflict with the neighbors.

Goats posed the greatest threat because they not only love the taste of leaves but also can climb trees to inflict damage to fruit-bearing branches. Due to these proclivities, keeping them was forbidden or limited in large parts of central Europe. In the Alps these restrictions were less extreme, however, and goats were allowed to join the roaming livestock if there was a herder to keep an eye on them. Goats that harmed fruit trees were subject to cruel punishments that sometimes resulted in their death. One particularly gruesome method of dispatching a marauding goat was to hang it by its horns from a forked branch. If there was bad blood between neighbors, goats were sometimes even driven into orchards purposefully to inflict harm on the property.

Pigs were likewise unwelcome among fruit trees because of their habit of churning up the ground and eating fruit that had fallen from the trees. If keeping them out of the orchard was not an option, they were often outfitted with a metal ring through the snout to keep them from digging up the soil.

A GOOD HARVEST WAS the reward for successful careful orchard maintenance but also presented a new set of problems. What should be done with all the delicious fruit that became ripe at the same time and might spoil before it could be eaten? Was it possible to ensure a supply of fruit in the winter? Less perishable alternatives to fresh fruit included compote, jelly made from juice, preserves in syrup, and fruit pickles in vinegar. Another option was fermenting fruit into wine that made the gray days of winter more bearable.

Some fruit such as apricots, peaches, and cherries were cooked and then preserved in *eau de vie*, a kind of clear fruit brandy. Many jam recipes existed as well, but some sound quite unusual from our perspective: they were salty or sweetened with fruit juice, cider, or honey. Toward the end of the seventeenth century, the trend toward sweet jam was fueled by a leap in sugar imports from the French Antilles and the French colony Saint-Domingue, now Santo Domingo in the Dominican Republic.

Some fruit was stored so that it could be enjoyed fresh a few months later. The rigorous work to make this possible took place on the periphery of the orchard. It was centered in the *fruiterie*, a building specifically constructed for storing fruit. The precious contents of a *fruiterie* were protected from damp, freezing tempera-tures, and hungry rodents. But maintaining such a building was a luxury available only to the wealthy who could afford the person-nel required. Until the middle of the twentieth century, the rest of the Northern Hemisphere's population had little choice but to go without fresh fruit during the winter months and try to make up for the lack of vitamins later.

For a precise description of how to store fruit, we can turn once again to La Quintinie. He explained that a *fruiterie* needs a strong wall on its north side to protect it from the especially cold winds from that direction. Insulated doors and windows are also

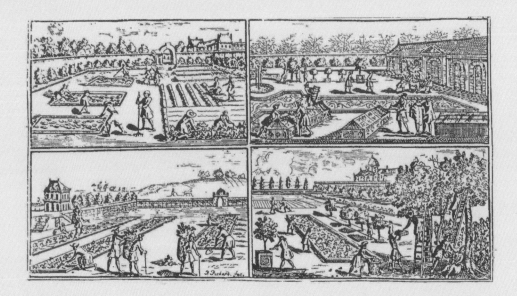

above
Four views of the
fruit and vegetable
gardens at Versailles,
La Quintinie, *The
Complete Gard'ner,*
1695.

a must. A large table in the middle of the room provides a spot for fruit baskets and porcelain plates. He suggested hanging slightly slanted shelves on the walls for holding the fruit, and attaching labels showing the type of fruit placed there and the date for eating it. The fruit should be carefully bedded on a layer of moss or fine sand that would absorb extra moisture. Workers need to check the room regularly, ensuring good ventilation and removing spoiled fruit immediately so that rot-inducing microbes don't spread. During especially cold periods, wood or coal stoves must be lit to keep the building warm. And of course, traps—or a cat—should keep mice and rats in check.

At this time, cookbooks still sometimes cited old dietary guidelines specifying that fresh figs, peaches, plums, apricots, blackberries, and cherries should only be eaten before a meal, while pears, apples, quinces, medlars, and sorbs should be eaten afterward. (Medlars and sorbs, lesser-known fruits today, are members of the Rosaceae family and somewhat resemble apples.) Otherwise, most fruit was used as a seasoning and cooked. But changes in attitudes toward fresh fruit were becoming apparent. One example appears in the 1784 handbook *L'école du jardin fruitier* (*The School of the Fruit Garden*), where a certain Monsieur de La Bretonnerie writes, "A fruit at full ripeness can only be healthy... even when fruit cannot cure illnesses, it can relieve their symptoms and even offer protection."

A NUMBER OF CHANGES have taken place at the garden of Versailles, but the geometrical structure is recognizable today, more than three hundred years after it was first laid out. And while some of the surrounding walls have been removed, the Sun King's pompous gate still stands. The thousands of plants growing there represent 450 types of fruit, such as 140 varieties of pear and 160 varieties of apples, including some from past centuries. Since 2007, Antoine Jacobsohn, an American trained at Cornell University and other institutions, has been the head gardener. His favorite pear is the buttery-juicy Duchesse d'Angoulême. Another botanical treasure to be found in the garden is Api jaune, a legendary apple known to Pliny the Elder.

— 8 —

Moving North

Bʏ ᴛʜᴇ ᴛʜɪʀᴛᴇᴇɴᴛʜ ᴄᴇɴᴛᴜʀʏ, fruit trees that had formerly been found in southern regions were definitely flourishing on the British Isles. During this period, fruit cultivation grew into a business. Records from this time show that in England's West Country, workers in the orchards received part of their pay in the form of hard cider—a practice that continued in this part of the country until the relatively recent past. In the fourteenth century, London markets were supplied by orchards found on Tower Hill, in Billingsgate and Eastcheap, and on Lombard Street and Bow Lane. Many authors of works on fruit farming packaged their practical tips in verse. For example, *Five Hundred Pointes of Good Husbandrie*, a horticultural guide by Thomas Tusser from 1580, contains the following melodious recommendations for the harvest:

Fruit gathered too timely will taste of the wood,
Will shrink and be bitter, and seldom prove good.
So fruit that is shaken, and beat off a tree,
With bruising in falling soon faulty will be.

As in France, fruit growing was an important component in royal gardens. In 1784, in the reign of George III, fruit specialist William Forsyth was appointed as the gardener at Kensington Palace. In addition to taking cuttings to start the soft-fruit gardens at the site, he expanded the western part of the property with a new orchard and beds for melons and cucumbers. Forsyth seems to have given up growing the vegetables that had been cultivated there to focus completely on fruit, and the harvest from Kensington joined the supply coming in from the other gardens. We don't know whether the resulting deliveries—mostly of peaches, but also including nectarines, apricots, grapes, raspberries, figs, plums, pears, and a single melon—were enough to satisfy the needs of their majesties and the rest of the royal household. In any case, there was nothing left over, and this had not always been the case when it came to the vegetables grown for the previous royal family.

When John Townsend Aiton was named the head gardener of Windsor Palace in 1804, he became the fruit supplier to the palaces of London. His fruit-growing efforts were spread over four separate kitchen gardens in Windsor park, each with its own pinery—a greenhouse for pineapples—and hothouses for accelerating the growth of grapevines. The gardens were located so far apart that Aiton needed half a day to ride to them all. His agreement with his royal clients specified that his job was

> to convey the produce in the most proper and commodious manner possible… to any place at which His Majesty or any part of the Royal Family or servants… may be resident, provided the distance doth not exceed 22 miles from Windsor Castle.

Furthermore, the men who transported the fruit to the other palaces were paid "head money": deliveries to Kew earned four shillings per day while those to St. James were compensated with five shillings. Despite a range of difficulties—which were not unique to Windsor—Aiton remained in his position until 1830, the year that William IV took the throne.

BRITISH GARDENERS HAD LONG paid careful attention to everything orchard-related in France, but it appears their success was mixed. Either that or it was a sense of national superiority that colored the impressions of François de La Rochefoucauld when he visited Suffolk in 1784. His assessment was certainly harsh:

> Kitchen-gardens are not as well-kept as ours: the gardeners are not so fully trained in their work. I've noticed that often

their trees were not well-pruned. They seem to like long branches that decorate the whole wall with leaves, and which naturally bear nothing like so much fruit as ours do. They are not familiar with the use of wire, and attach each branch with a piece of cloth and a nail. In general, all that they know about the cultivation of kitchen-gardens and the various kinds of fruit they have, comes from France.

Fruit needs heat to ripen. And especially in the climate of Britain, heat can be hard to come by. One solution to this problem was encircling large spaces with a wall so that fruit trees could be grown in climate-controlled conditions. A wall around a kitchen garden had to be high enough to protect the plants from the elements but not so high that it completely blocked the sun. Six to ten feet (about two to three meters) was typical. Walls were made of stone, plaster, and earth and sometimes topped with a tile roof, which offered some protection from the rain: water was less able to seep in and the wall would last longer. Sealing the wall with mortar or plaster, in turn, helped to keep out rodents and insects. Of course, a wall was also a sign of ownership. At a time when people were considerably shorter than they are today, a six- to ten-foot wall was also effective at keeping intruders out. And while the walls themselves created warmer conditions, some had the added advantage of being heated. At the West Dean Estate in Sussex, for example, the head gardener's cottage was built against one of the walls, lending the wall some heat. In addition, walls were sometimes built with two layers of brick and a space in the middle. Slow-burning straw could be packed into these walls and set alight.

Gardens and orchards surrounded by massive stone walls were a common sight throughout Great Britain and other European countries that shared a similarly chilly climate, such as the Netherlands, Belgium, and northern France. The walls not only protected the plants from the cold north winds but also absorbed the sun's heat during the day and released it during the night. As a result, they could create microclimates that were 18 degrees Fahrenheit (10 degrees Celsius) warmer than the surrounding area. These conditions nearly matched those much farther south at the Mediterranean. In the sixteenth century—during the Little Ice Age—the Swiss botanist Conrad Gessner described the positive effects of such walls on the ripening process of figs and currants.

THE SUCCESS IN GROWING plants from southern Europe kindled the desire for other potential imports from faraway lands. Gardeners like John Tradescant the Elder (ca. 1570–1638) undertook adventurous journeys through Europe and North Africa in search of unknown or rare fruit and ornamental plants. On the Barbary Coast, the home of the Berbers in North Africa, he discovered the Petit Muscat or White Algiers apricot. But Tradescant's greatest love appears to have been plums: *Tradescants' Orchard*, a collection of luminous watercolors attributed to this great gardener, records twenty-three different varieties of them. He especially appreciated their varying shapes and colors—not surprising, perhaps, when we realize that he had lost his sense of smell. Although his book existed only in manuscript form and was never printed, it has gone down in history as one of the works inspired by the myth of the fruit goddess Pomona.

left and above
Pomegranates and a walnut tree from *Helmingham Herbal and Bestiary*, early sixteenth century.

In 1630, Tradescant entered the service of King Charles I as the supervisor of the gardens, vineyards, and silkworms at Oatlands Palace in Surrey. He and his son, also called John, achieved fame for the Ark, their cabinet of curiosities in south London. They cultivated their botanical discoveries on the grounds. The collection of artifacts gathered while father and, later, son were scouring the world for botanical specimens was the foundation for the first public museum in Britain, the Ashmolean Museum in Oxford.

THE THOUGHT OF SCOTLAND conjures up images of a raw climate and wild nature rather than blossoming orchards and ripening fruit. But the country has a few surprises in store. Thanks to the Gulf Stream, the weather on the western coast is relatively mild. The region's first orchards were planted in the twelfth century by the religious orders there, including the Dominicans or Black-friars, and probably resembled the cloister gardens of central Europe. Records from Paisley Abbey, near Glasgow, which was founded in 1163, show that the complex included a six-acre (two-and-a-half hectare) orchard. A royal orchard existed in Edinburgh as well. It was first documented by the English in the 1330s, but it had likely existed since the time of David I, who ruled Scotland from 1124 to 1153. The large expanse of gardens and orchards surrounded the southern and western flanks of Castle Rock. The area provided fruit and produce for the palace and might well have been used commercially, as well. There are also records of timber being cut from the orchard, indicating that it provided more than just fruit.

The first precise information about the fruit grown in Scotland comes from the late seventeenth century, when William Lindsay, the eighteenth Earl of Crawford, made a list: twenty-six types of apples, forty types of pears, thirty-six types of plums, twenty-eight

left
A cornucopia of pears, Johann Hermann Knoop, *Pomologia*, 1760.

types of cherries, various varieties of peaches, nectarines, and apricots, and gooseberries and currants. It is safe to assume that the peach and apricot trees needed a protected spot by a wall in the sun. Scotland's largest orchards could be found at Hamilton Palace by the River Clyde near Glasgow. A 1668 visitor to the garden recorded the following impressions:

> *Great abundance of as good vines, peaches, apricots, figs, walnuts, chestnuts, philberts etc. in it as in any part of France: excellent Bon Crestien pears... The walls are built of brick, which conduces much to the ripening of the fruits.*

In her book *The Days of Duchess Anne*, Rosalind Marshall describes how the third Duke of Hamilton, William Douglas-Hamilton, continued the tradition of fruit growing started by his ancestors. He hired brickmakers to produce thousands of bricks for new walls so he could espalier peaches, apricots, and cherries against them. He had great success, and in 1682, the Earl of Callander was eager to discover "at what distance My Lord Duke plants the trees on his walls." He sourced pear and cherry trees locally and had peaches and apricots sent up from London. He also procured grapevines, mulberries, and nut trees from south of the border. To obtain grafted trees, which were popular at the time, he extended his reach as far as the continent, on one occasion sending a Bo'ness shipmaster to Holland for "5 peaches upon apricots, 4 peaches upon plums and 2 apricots."

Gardeners trying to create the optimal conditions for the fruit to ripen sometimes took the wrong turn. For example, they understood early on that black surfaces absorb heat most easily. But when they painted the walls behind their peach espaliers black, the new shoots developed too soon and froze as soon as the temperature dipped. In summer, the same walls sometimes grew hot enough to burn the delicate fruit.

In the British Isles' walled gardens, fig trees were first trained into fan shapes or espaliers. Around the mid-nineteenth century, however, gardeners began building special glasshouses for these plants. One advantage of this approach was the ability to harvest figs twice or even three times per year. One unusual architectonic solution can be found in the garden of Ardgillan Castle in Ireland: a wall that winds to create twenty alcoves offering additional protection from the cold and wind. Most likely, these niches were used to house especially sensitive

nectarine, peach, and pear trees. The idea caught on across the Atlantic: Thomas Jefferson was so taken with it that he had similar serpentine or "crinkle-crankle" walls built at the University of Virginia in Charlotte, which he founded. They separate the gardens surrounding each of the ten pavilions on the university's main lawn.

While some types of fruit were especially suited to walled gardens, all kinds of considerations could influence what was planted in them. In the area around Nunnington Hall, near York, orchardists focused on apples that could be stored for long periods and were therefore suitable to supply British sailing ships. They included varieties with the wonderful names Dog's Snout (with a quince-like shape that indeed resembles a dog's nose), Cockpit, and Burr Knot.

Walled fruit gardens were also planted in northern France, and many still exist there. They were especially common in Montreuil-sous-Bois, a town on the eastern edge of Paris famous for its peaches. At the high point of peach cultivation in the 1870s, more than 370 miles (600 kilometers) of wall crisscrossed the area. The presence of these barriers must have influenced every aspect of life. It's not hard to imagine how the children would have clambered over and along such an obstacle course. For outsiders, the resulting labyrinth was nearly impenetrable. When the Prussians occupied Paris in 1870, the army supposedly took a wide detour around Montreuil rather than risk getting lost there. While they no longer serve their original purpose, the walls have proven to be astonishingly resistant to the pressures of urbanization. Some of them stand to this day, structuring the landscape and marking property boundaries.

below
Fruit walls in Montreuil-sous-Bois near Paris, early twentieth century.

— 9 —

Orchards for the Masses

THE ORCHARDS OF PAST times that have survived—in physical form or at least recorded in written documents—are those that belonged to kings and queens, the noble class, and religious orders. But while these estates were impressive, they did not represent the typical orchards that provided most people with fruit. As far back as the Middle Ages, networks of orchards surrounded many of central Europe's villages and towns. In many cases they were part of or located near vegetable or kitchen gardens. Crops such as potatoes, turnips, corn, or grasses grew in the shade of the trees and both crops and fruit were harvested by hand in the fall.

In the past, boundaries between the orchards planted near settlements and surrounding forests were often fluid. People went out into the forest to gather fruits and nuts, but they also used the forest as a source for rootstocks that they could then use to graft on scions from cultivated trees that were more productive than those growing wild. Information on the region around Paris in the eighteenth century provides examples. Wild cherry trees were collected as rootstocks for cultivated cherries. Hazelnut bushes and wild apple and pear trees were useful as rootstocks, too. And the currant bushes growing in the forest could simply be transplanted—no grafting required. As forests were often much drier than cultivated orchards and gardens, it often took some time for forest rootstocks (or sometimes transplanted bushes) to acclimate to their new surroundings.

previous page
The Cider Orchard
by Robert Walker
Macbeth, 1890.

It is also important to remember that walnut and chestnut trees, often cultivated during the time of the Roman occupation, ran wild and multiplied, becoming part of the wider landscape's hedges and forests without any human interference. Forests were not only a source for fruits, nuts, and rootstocks; they were also a source of packing material for fruit to be transported, especially the delicate cherries. Chestnut leaves cushioned the precious cargo from bumping and jostling on the way to market. In Britain, ferns helped to protect damage-prone fruit such as summer pears. Fruit baskets, known as *maunds* (the name is an Anglicized spelling of a measure used in the Mughal and Ottoman empires), were lined with the soft leaves. Packing the fruit with the stems facing out ensured that pears wouldn't poke holes in their neighbors.

In the second half of the eighteenth century, people in some places were collecting so many oak, hazelnut, and chestnut leaves that the authorities were compelled to pass laws reining them in. And they continued to gather wild fruit and nuts, including hazelnuts and chestnuts.

All these fruit and nut trees scattered across the countryside defined the landscape in physical, symbolic, and aesthetic terms,

lending it an important part of its character. They fundamentally changed the environment, adding depth and beauty to the surroundings. Just think about how high these trees can grow: an apple tree can be thirty feet tall (ten meters), pear trees reach heights of fifty feet (fifteen meters), and cherry trees can stretch sixty-five feet (twenty meters) into the sky. And trees endure for generations, especially nut-bearing species. They can easily live a hundred years and some survive for five hundred—assuming they aren't cut down for their highly prized wood with its beautiful grain.

above
A German fruit seller with apples, plums, pears, peaches, and medlars, ca. 1840.

Henri-Louis Duhamel de Monceau, who made an appearance earlier, generally wrote about highly cultivated and trained espaliers. But in a treatise from 1768, even he wrote that the fruit "of freely growing trees is superior to all others." A claim like this from such a celebrated expert must have helped the image of naturally growing trees at a time when gardeners found the symmetrical espaliers more attractive. And Monceau's praise didn't end there:

If a regular form is desired, nothing more is required of this tree than the removal of the dead wood and a few branches ... Left to the care and direction of nature, it will stretch its twigs and roots in all directions. Its sap flows strong and plentiful to the outermost ends, strengthening and multiplying the twigs that are essential to the tree's growth and vigor.

The importance of fruit farming was recognized beyond France and Great Britain. In the German-speaking parts of Europe, the highest authorities encouraged progress in the cultivation of fruit trees. The earliest documents from this area about fruit varieties were the previously mentioned *Capitulare de Villis* issued by Charlemagne himself. They constitute the first laws of the Middle Ages to regulate land use and agriculture, and were intended, among other things, to ensure a good supply of fruit throughout the kingdom. Administrators had realized how beneficial fruit could be in preventing famine and wanted to be prepared. The recommendations laid out in this legislation cover sixteen species of trees, most of them fruit trees. Efforts to incentivize fruit farming continued long after Charlemagne. Under one sixteenth-century law, for example, every couple was required to plant and care for six fruit trees—otherwise they were not allowed to marry. Another good reason for planting fruit trees was that the fruit could be processed into beverages. Primitive or nonexistent sanitation and activities like tanning and lead production left drinking water increasingly polluted, and dirty water was believed to cause plagues. It's no wonder that demand for juice, beer, and wine increased.

Like their French and British counterparts, German fruit farmers drew on resources from the forests. ("German" is used here to indicate people living across a large number of kingdoms and districts who spoke German and shared a broadly similar "German" culture—Germany as a political unit did not exist until 1871.) In an edict from 1567, for example, Christoph, Duke of Württemberg, permitted his subjects to dig up young wild-growing fruit trees. It therefore seems wild fruit trees could be found growing throughout the landscape and were an alternative to grafted trees, which were unavailable in sufficient amounts from commercial nurseries or too expensive for most people to purchase.

left
The Bavarian town of Bamberg, set in a landscape of vineyards and orchards, 1837.

Shortly after taking the throne in 1740, Frederick the Great, the absolute monarch of the major state of Prussia, informed his provincial authorities that "the cultivation of fruit trees is to be encouraged throughout the land wherever practicable." He was responding to the fact that the Thirty Years' War (1618–1648) had laid waste to many of central Europe's orchards. At the same time, he wanted to ensure that his soldiers could access supplies of fruit on their maneuvers through his territory. (Surely many others, such as travelers, also benefited through the ensuing ages whenever fruit trees growing somewhere in the countryside were replaced.)

Yet these royal efforts apparently were not as effective as Frederick had hoped—just three years later he issued another edict imposing a fine on towns and provinces

> *for every three-score fruit trees, willows, lindens etc. unplanted that, in the judgement of the land and tax administrators, they could have planted.*

And this was not the last such penalty Frederick threatened. Year after year, he demanded precise information about new fruit trees and those that had died. But it still wasn't enough. In 1752, a few years before the Seven Years' War began, His Majesty ordered:

> *Good, communal nurseries should be established in all villages and a man knowledgeable in raising trees should be engaged who can also provide instruction to the inhabitants... Every farmer must plant at least 10 young fruit trees each year. The excess fruit must be baked and offered for sale to the townspeople.*

The king's instructions to bake the fruit did not refer to cakes or pies, but rather to a process of drying the fruit in order to preserve it. It was no coincidence that bakehouses for this purpose could often be found near large orchards. In the town of Treffurt in Thuringia (central Germany), the business of fruit became a mainstay of the local economy. As the bounty of fruit grew, so did the ring of drying kilns around the town, where cherries, plums, pears, and apples were turned into marketable, durable goods; in the fall they bustled with life and the pleasant scent of fruit spread around them. The fruit was cut into pieces before the drying process and the end product was normally eaten with lard or pastry.

Yet the official commandments to plant fruit trees did not always have the desired effect, and in some cases they weren't even taken seriously. A 1765 report from Silesia, now part of Poland, complains that when inspectors made their rounds "mere cudgels or rods were stuck in the earth" to create the illusion of young trees growing there.

Furthermore, nursery operators were the targets of a great deal of mistrust. Quality standards were nonexistent until well into the nineteenth century. Buyers of young grafted plants always had to be prepared for the possibility that their purchases would either not develop properly or produce a different kind of fruit than what they had wanted. And the use of inconsistent names for all the different varieties was a breeding ground for confusion.

Shortly before the start of the nineteenth century, the ducal nursery at Solitude Palace, near Stuttgart, was selling an impressive one hundred thousand plants each year. The nursery was run by the former barber Johann Caspar Schiller. While his son was the famous poet Friedrich Schiller, father Johann Caspar was an author in his own right. His book, *Die Baumzucht im Großen (Large-Scale Tree Cultivation)*, contains helpful lists such as "trees that are

highly suitable for lining roads" and "trees that grow quickly and strongly." These topics line up well with developments during the early nineteenth century, when tall-growing fruit trees were planted like never before, particularly in the southwestern parts of German-speaking Europe. As a result, they ringed the towns and villages, bordered the streets, and took up position on the pastures and farmland, sometimes in the middle of potatoes and other "fruits of the field." Areas on sloping ground unsuitable for planting crops were especially popular locations for trees. In addition to fruit trees, many nut trees were planted during this period: their wood and oil were in demand.

A certain Professor Hopf, who traveled through the Erms river valley south of Stuttgart in 1797, recorded his impressions of a "fruit forest" he encountered there. According to his account, it

above
Carloads of fruit arriving at Stuttgart train station, 1899.

stretched for several miles and "the harvest was turned into cider, dried fruit, or brandy." A popular saying at the time captures just how much people depended on the nearby trees: "Plant a tree in each spot you see, and keep it fit—you'll benefit!"

The tall-growing fruit trees not only added variety to the landscape but also brought real ecological benefits. As a horticultural pioneer, Johann Caspar Schiller was quite aware of them:

> Those who practice the cultivation of trees obtain a pleas-
> ant share of their sustenance from it. It suffices to adorn the
> countryside, purify the air, provide protection and shade, and
> most excellently serves the needs, pleasure, and comforts of life
> for both man and beast.

It was surely no coincidence that into the eighteenth century, fruit farming was especially widespread in regions such as the south of England and the southwestern German territories—places where it was possible to draw a direct line back to the period of Roman occupation. During the eighteenth and nineteenth centuries, large orchards were part and parcel of the landscape there, especially in those areas with a favorable climate. A 1908 report from Alsace, the famous wine region between the Vosges mountains and the Rhine that is now part of France, gives a sense of what they were like:

> Generally, the orchards are located around the village or up on
> the flatter part of the vine-covered hillside, and lend an enchant-
> ing touch to the appearance of the place. The "fruit fields," which
> are usually divided into very small plots, have existed here since
> time immemorial, and were even more plentiful in earlier cen-
> turies than they are today.

left
Apples for all,
Germany, late nine-
teenth century.

Fruit trees got another boost in the nineteenth century when
the grape phylloxera, a destructive insect closely related to aphids,
hitched a ride from America and laid waste to vast swaths of
Europe's wine country. In many cases, when the vineyards disappeared, fruit trees were planted in their stead.

Fruit trees in the countryside—whether growing wild in the
forest or cultivated by villagers—also played an important role for
another group of people: the inhabitants of the constantly growing cities. Unfortunately, when they were far enough from home, it
was easy for them to imagine that the rules no longer applied and
they could help themselves to the bounty growing on the boughs.
The British historian Richard Cobb describes the lawlessness in
regions around Paris. "It was highly dangerous, to say the least of
it, to walk or ride on any of the highroads at dusk and during the

night," he explains. "At dusk, the passerby might be confronted by silently moving lines of shadowy figures, their backs bent under the weight of trunks and piled-up wood, as they headed for home."

Cobb is describing people illegally cutting trees for firewood, but fruit and mushroom thieves were just as common. In the fall, desperate people often resorted to nutting and blackberrying, especially during wartime. For rural communities, these plundering strangers, driven by hunger and cold, were invaders who damaged crops and left gates open so that farmers risked losing their animals. At times, however, they also found their way into places uncrossed by fences or hedges that no one had really laid claim to. Seen in this light, they were simply taking advantage of the freedom available and profiting from the absence of authority—and of course from whatever wood, fruit, and mushrooms they could find. Described in verse, it all sounds quite romantic, as in "The Shepherd's Calendar," a poem from 1820. (The author, John Clare, lived in Helpston, a village in the English fens):

The bawling song of solitary boys,
Journeying in rapture o'er their dreaming joys,
Haunting the hedges in their reveries,
For wilding fruit that shines upon the trees.

At another point, Clare writes:

Still to the woods the hungry boys repair;
Brushing the long dead grass with anxious feet,
While round their heads the stirr'd boughs patter down,
To seek the bramble's jet-fruit, lushy sweet,—
Or climbing service-berries ripe and brown.

The list of ways that fruit trees positively influence their surroundings goes beyond sustenance, shade, and an attractive landscape. Today we know that in meadows, trees inhibit evaporation and increase moisture in the soil. The result is a microclimate that resembles the forest. Under these conditions, the growth beneath the trees provides a welcoming home for all manner of small animals and plants. Many bird species find food there and can nest in spots such as holes in trunks and branches. If the meadows are farmed carefully or left largely alone, rare plants like orchids may even take up residence.

Furthermore, tree roots hold the soil and protect it from erosion—a benefit that is especially useful on hillsides, where fruit trees were often planted. And fruit or other deciduous trees are ideal neighbors for houses because they provide shade during the summer but let in sought-after light in the winter when their branches are bare. Trees also reduce the risk of damage from the weather: the famous hipped-roofed houses of the Black Forest were often protected by a row of "house trees"—often walnut trees—on the side where they would block the prevailing winds. For centuries, homesteads and fruit groves, wild or cultivated, went hand in hand.

Here and there the gnarled trees of traditional fruit farming can still be found, but in most cases they have been swept away by the utilitarian demands of our modern age. The rows of fruit trees that once lined roads and pathways have fallen victim to standards of traffic safety: a car running into a tree often has deadly consequences.

In the Dordogne valley in southern France, in an area known as the Périgord, an entire landscape of broadly spreading oak, chestnut, and walnut trees has existed for centuries. Périgord nuts have long been this region's pride. There used to be even more trees—so

many that people claimed squirrels could cross the entire Périgord simply by jumping from one to the next. The Route de la Noix du Périgord through the area is a testament to a time—from the Middle Ages to the Renaissance—when nuts were used as currency and accepted for purposes from settling debts to paying customs duties. People harvested the nuts by beating the branches of the trees with long poles. And reportedly the villages echoed with the sound of small iron hammers breaking open the shells.

The remaining nut farmers face tremendous competition from California, China, Iran, and Chile. Staying in business requires creativity: they feed the nuts to ducks and geese so that their livers grow fat and rich for pâté; they distill the nuts into walnut liquor or extract their oil. Nuts are baked into bread, processed into sweet spreads, incorporated into pesto or sausage. With a little bit of luck, you might find one of the old oil mills among the farms, wreathed with rose bushes, that dot the romantic landscape. The mills' wooden wheels are powered by water, the shells burst under hydraulic hammers, and the nutmeats dry in red-hot ovens. It takes more than thirteen pounds (six kilograms) of nuts to make a little over thirty fluid ounces (a single liter) of oil.

— 10 —

Cherry Picking

CHERRIES ARE A SYMBOL of eternal, paradisiacal spring and of erotic temptation and seduction. Is there any other fruit that looks so much like lips puckered up for a kiss? To say nothing of the wonderful aroma and taste. It's not at all surprising that cherries have long been objects of desire.

In the time of Pliny many cherry varieties were already being cultivated. Although we have no way of knowing exactly what they were like or whether they were sweet or sour, some of their evocative names have survived. Aprionian was said to be the reddest and Lutatian the darkest. Caecilian was especially round and Junian tasted particularly good when eaten straight from the tree. But the best of them all was the celebrated late-maturing Duracina, known for its dark color, juiciness, and relatively hard flesh. In more recent times it was also called the "heart-cherry." A variant, Tarcento Duracina, enjoyed wide popularity in the nineteenth century and can still be found occasionally in the northeast of Italy, near Udine.

Such cultivated cherries were not native to Rome, however, and their arrival in the city was the stuff of legend. In 70 BCE the Roman general Licinius Lucullus conquered lands around the Black Sea from the famous king Mithridates VI, including a city that the Romans named Cerasus (today Giresun in northeastern Turkey) after the precious cherries, *cerasia* in Latin, that grew there. Supposedly, when Lucullus celebrated his victory with a triumphal march through Rome, cherry trees were among the booty on display. He then planted them in his garden, where they bore fruit. Cherries were served as a fitting culmination of great feasts but were also dried or preserved in honey. Cherry juice provided the basis for fruit wine as well.

In the eighteenth century, cherries had a special place at the court of Prussia's rulers in Potsdam. The royal gardens often featured large numbers of cherry trees and an entire culture developed around the fruit. Frederick the Great had a passion for cherries that seems nearly libidinous. In 1737—as a young man of twenty-five—he confessed his longing in a letter to a friend:

> On the 25th I will travel to Amalthea, my precious garden in
> Ruppin, and I am burning with impatience to see my vineyard,
> my cherries, and my melons once again.

The cultivation efforts in the king's gardens aimed to ensure a supply of fresh cherries over the longest possible period. Varieties that ripened either early or late were therefore in demand. "Forcing" also played a role: trees were planted against southern-facing walls where the sunlight (if the sun appeared at all during the short winter days) was especially strong. Glass panels placed

at an angle in front of these walls intensified the sun's rays and helped keep the trees from cooling down at night. At the king's orders, cherry trees also dominated the kitchen garden to the west of Potsdam's city wall—there were 328 of them, along with a 260-foot-long (80-meter) greenhouse for nurturing the young plants. Cherry forcing reached its zenith in the form of special "cherry boxes" for small plants that were kept warm with horse manure. As unlikely as it seems, small numbers of cherries were supposedly harvested as early as December and January and sold by the king for the princely price of two thalers each. Trees were available in five formats: standard, semi-dwarf, dwarf, pyramids, and espaliers. A special breed was even developed to be used in hedges.

below
A couple reaching for cherries, Germany, 1616.

A list from the pharmacy for Berlin's palaces and royal court from 1758 shows that in addition to being eaten fresh, the king's cherries found their way into a wide range of products: cherry-lavender water, black cherry brandy, sour cherry syrup (with and without marigold blossoms), and sour cherry preserves. In 1764 the poet Anna Louisa Karsch, celebrated by some of her contemporaries as "the German Sappho," wrote "In Praise of Black Cherries" as a paean to the fruit:

A host of bards have raised their song
To praise the grapevine's bursting jewels
Why then does not one come along
To sing the cherry's virtues?

Once did this ruby globe amid
The lovely boughs of Eden ripen
And into great temptation bid
Milton's fair heroine.
...

A toast! I raise my glass three times!
To laud the rose is common practice.
Ye poets, gather up your rhymes
And praise the cherry's blackness!

In central European folklore, cherry trees are often associated with the moon. Those who ventured under them during a full moon risked encountering spirits who were up to no good. It was dangerous to even secretly observe the elves and fairies dancing under the moonlit cherry blossoms.

Ripe cherries must be eaten or processed right away. And they have always led pickers into temptation. "Two in the basket, one in the mouth" was one typical rule of thumb. Those who took things too far often paid the price in stomach pains. When the cherries were ripe, it was essential to not only work together to pick them quickly, but also get them to market. The carts set off for town on the evening of the harvest and the horses traveled through the night. While cherries were extremely popular, people did not always appreciate the naturally growing trees. Because they grow up to sixty-five feet (twenty meters) tall, pickers needed to be good climbers in order to reach the fruit before it simply rotted on the branches.

No region of England is as synonymous with cherries as Kent. The region's cherry tradition originated during the reign of Henry VIII, who ordered an orchard to be planted at the town of Sittingbourne. Now one of the best places to enjoy the trees—especially when they bloom extravagantly in the spring—is the National Fruit Collection at the Kentish city of Brogdale. The largest such collection in the world, it features 285 types of cherry trees and hundreds of examples of other species, including apples, pears, plums, currants, quinces, and medlars.

But Kent is not the only place where delicious cherries grow. The area around Lake Constance, which lies at the point where Germany, Austria, and Switzerland come together, is also a paradise for fruit trees with a farming tradition that stretches back for countless generations. Near the quiet little city of Ravensburg, Joachim Arnegger farms a seven-and-a-half-acre (three-hectare) cherry orchard. Yet despite the good conditions in the region, he faces a challenge: cherries bloom for just two to three weeks, and the temperatures during this time are often still cold enough to discourage the honeybees from emerging from their warm hives.

To ensure a good harvest, Arnegger brings over hornfaced mason bees from neighboring Switzerland to pollinate his orchard. These "temporary workers," as he calls them, arrive in groups of five hundred packed in nest boxes. For a fruit farmer, the hornfaced mason bees have some clear benefits over their honeybee cousins. For one thing, they are much less picky about the weather, emerging for work as soon as the temperature climbs to about 40 degrees Fahrenheit (5 degrees Celsius.) Even a bit of rain doesn't stop them. And they are downright obsessive: up to three hundred times more effective than honeybees. This relentlessness is due to the evolutionary programming that drives them to breed and raise their offspring in the short span of just four

above
Gathering fruit from
a hedgerow, late nine-
teenth century.

weeks. Another advantage is that hornfaced mason bees switch from one blossom to another more often than honeybees, which leads to more mixing in the fruit trees' gene pool. What's more, nearby fields of rapeseed bursting with fragrant yellow flowers don't interest them at all. They stick to the fruit trees. In the fall, the farmers send the nest boxes back to their owners. The young brood inside is subjected to a special treatment that kills mites and fungus and then stored in a refrigerator for the winter.

Although renting the bees is expensive, the procedure pays off for Arnegger. With the mason bees' help, his trees can produce twenty-two tons of cherries in a good season. He sells this red bounty entirely on his own, without even the support of a cooperative. If you look closely at his orchard, you notice that stinging nettles and all manner of other weeds grow along its borders. Dead branches also lie wherever they fall, providing a home for beetles and other creatures. Although Arnegger does not fit the classical description of an "organic" farmer, evidence abounds at

below
Illustration on the
cover of a Turkish
magazine, ca. 1925.

his farm that he understands the interconnections in the system of nature. And the idea of using the highly active mason bees to pollinate orchards is catching on. In Japan, farmers are increasingly turning to these industrious insects for help. And their assistance is urgently needed: invasive mites have decimated the country's honeybee population. In nearby China, so many insects have disappeared that the fruit trees must be pollinated by hand.

Of course, unpollinated cherry trees would be more than just a culinary disaster in Japan. The tiny, delicate cherry blossom or *sakura* is one of the most important symbols of Japanese culture, and the cherry blossom festival is a highlight of the year. This pleasure is short-lived—most varieties bloom for just a few days— but for that reason it is all the more intense. One popular activity is gathering for *hanami* (cherry-blossom viewing) parties, where friends, family, or colleagues sit together to socialize and enjoy the view. The timing of the blossom season varies from region to region, starting as early as January on the warm subtropical island of Okinawa in the south and progressing northward from there over the following months.

Humans are not the only beings who are seduced by the cherry trees' luscious fruits: the little glowing red globes are a powerful lure for all kinds of birds. To some gardeners, these winged guests are nothing but pests, but others enjoy the music and merry-making of their feathered friends at a cherry feast. One was the English writer Joseph Addison, who vividly described the delights of such an experience in a 1712 article for the *Spectator*:

> There is another circumstance in which I am very particular, or, as my neighbours call me, very whimsical: as my garden invites into it all the birds of the country, by offering them the conveniency of springs and shades, solitude and shelter, I do not suffer any one to destroy their nests in the spring, or drive them from their usual haunts in fruit-time. I value my garden more for being full of blackbirds than cherries, and very frankly give them fruit for their songs. By this means I have always the musick of the season in its perfection, and am highly delighted to see the jay or the thrush hopping about my walks, and shooting before my eye across the several little glades and alleys that I pass through.

Le Pont St Louis

— 11 —

Pucker Up

ITALY HAS LONG BEEN synonymous with citrus groves. But how did the country achieve this status? Lemon trees were cultivated early on in Persia, and legend has it that Alexander the Great brought them back with him around 300 BCE. A hundred years later, Greek settlers took them along to Palestine. Soon they were growing in Italy as well, but the trees were destroyed by invaders and survived only on the major islands of Corsica, Sardinia, and Sicily. Their further spread was due to the fact that the Arabs brought lemon and orange trees to Andalusia, in the south of the Iberian Peninsula, along with olive trees and—despite the prohibition on alcohol in the Quran—wine grapes.

Do you know the land where the lemon-trees grow,
in darkened leaves the gold-oranges glow,
a soft wind blows from the pure blue sky,

the myrtle stands mute, and the bay-tree high?
Do you know it well?
It's there I'd be gone,
to be there with you, O, my beloved one!

It's easy to assume that the German poet Johann Wolfgang von Goethe (1749–1832) wrote these famous lines under the influence of his legendary journey to Italy, but in fact he composed them three years earlier in the decidedly un-Italian town of Weimar. The nearby Belvedere Palace boasted a baroque orangery with two pavilions and an enclosed courtyard. Fragrant lemon and orange trees grew in the greenhouses in pots so that they could be moved outside in the summer.

While they could hardly compare to the vast groves of Italy, collections of citrus trees were painstakingly grown at a number of places in Europe, especially palaces and other noble residences. And they inspired flights of fancy in the people fortunate enough to spend time around them. Jean-Jacques Rousseau had fond memories of an apparently quite productive stay at the Petit Château of Montmorency, about thirty miles (fifty kilometers) north of Paris:

> In this profound and delicious solitude, in the midst of the woods, the singing of birds of every kind, and the perfume of orange flowers, I composed, in a continual ecstasy, the fifth book of Emilius, the coloring which I owe in a great measure to the lively impression I received from the place I inhabited.

Of course, the previously mentioned compendium by La Quintinie includes his thoughts on cultivating orange trees—the specialty of the *jardinier orangiste* or orange gardener. The master gardener's words are a paean to these plants:

Pucker Up

In the entire garden there are neither plants nor trees that provide such pleasure and over such a long time. Not a single day goes by in the year without the orange trees providing something that can and must delight their devotee, whether it be the green of their lovely foliage, the gracefulness of their characteristic form, the profusion and fragrance of their blossoms, or finally the beauty, quality, and long availability of their fruit, etc. I admit that no one could be more delighted by it all than I am.

below
Differently shaped
lemons as observed
by Italian naturalist
Ulisse Aldrovandi
(1522–1605).

The seventy-two-page tract is bursting with precise instructions. The shape of the crown, for example, should resemble "the
form of a freshly closed mushroom or a skullcap," and the crown
should be full "without its interior being disorderly and jum-
bled." Achieving the "consummate beauty of the orange tree" also
requires "that it be free of all types of unpleasant odors, dust, and
infestations of insects such as aphids and ants."
And how exactly should orange trees be arranged?

*If one has a large enough greenhouse to accommodate two rows
of orange trees and positions them with taste and symmetry,
which can take a number of forms, one should do it in such a
manner as to leave a path free in the middle so that it is possible
to enjoy the beauty of the trees brought inside while walking.*

La Quintinie's musings were the blueprint for nearly every
French garden book written in the eighteenth century.
The lemon tree, which can bloom and bear fruit at the same
time, became a symbol for the Mediterranean region despite the
fact that it originated elsewhere. Did Goethe ever ask himself who
brought the lemon trees he celebrated so memorably to Italy?
Presumably he knew the trees originated with stock the Arabs
brought with them to the Mediterranean region. From the report
of Ibn Hawqal, a trader from Baghdad, we know that a number of
gardens with irrigation canals like those in Mesopotamia existed
near Palermo in the tenth century. At that time, the plants raised
there already included orange and lemon trees.
As we know today, citrus plants first emerged far to the east
of Italy. One piece of evidence is an observation recorded twelve
hundred years ago by the Chinese lyricist Du Fu (or, as his name
is sometimes transcribed, Tu Fu): "An autumn day, pavilion in the

wilds, a thousand orange trees fragrant…" At another point he refers to the singing of the so-called orange ode, which proclaims that orange trees should be kept in the homeland and not planted abroad. Someone clearly broke this rule, of course, and now oranges grow in tropical and subtropical locations around the world.

In *The Land Where Lemons Grow*, her wonderful book about citrus cultivation in Italy, Helena Attlee explains why it's difficult to determine exactly when the first sweet oranges arrived in Europe. Our sense of taste is subjective, and in the period in question it was not possible to measure the acid and sugar content of fruit. However, a consensus exists that seafarers brought sweet oranges to Portugal from China toward the mid-sixteenth century. Until that time, Europeans would have only known bitter oranges.

right
*Spring Evening Banquet
at the Peach and Pear
Blossom Garden*, Leng
Mei, China, ca. 1700.

It is no accident that the first known books on the study of fruit—a discipline that would later be known as "pomology"— come from China as well. They were written by Han Yen-Chih, who was the prefect of Wenzhou, a city in the province of Zhejiang. His *Chü lu* (1178) is a treasury of advice on breeding and grafting Ni-shan oranges, which scholars today believe were mandarins. Some of this information should be taken with a grain of salt: at one point, for example, Han Yen-Chih claims that the oranges will spoil quickly if the person picking them has drunk wine on the same day. But his descriptions of how oranges were used are quite illuminating. He describes orange perfume for clothing, orange flavoring in meals, and oranges preserved in honey. The peel with its essential oils was prized by traders and used in a variety of medications. The blossoms of one variety were processed into a powder that could be burned to release an orange scent. Contemporary accounts related that this fragrance was so intense it created the illusion—at least at the olfactory level—of sitting under

a real orange tree. Oranges were so important in ancient China that the court had an orange minister charged with ensuring an adequate supply of the fruit.

A wealth of garden-themed paintings and calligraphy from the following centuries has survived, often featuring imaginatively designed landscapes shown in exquisite detail. But despite all my research, I have found very few works that explicitly depict orchards. Furthermore, when fruit trees do appear it's not always clear what type they are.

One example of these challenges comes from the period around 1600, the golden age of the classical Chinese garden. The province Jiangnan, below the lower reaches of the Yangtze river, was once home to hundreds of gardens, both large and small. They included the Zhi Garden or "Garden of Repose" located to the north of the wall surrounding the city of Wujin. In 1627, the painter Zhang Hong, known for his monumental landscapes and topographical pictures, created a series of views of this garden. The fourteenth (of twenty) shows the "House of the Pear Cloud," a large pavilion with an elegant hipped roof and a terrace that overlooks a pond and provides a view of the moon.

Surprisingly, the pavilion is not surrounded by pear trees, but by plum trees in full bloom. Seven hundred of them are said to have grown there. Another image shows orange trees mixed in with magnolias, cypresses, and other plants. The distinction between the various species of trees is not evident from the paintings themselves: it is provided in accompanying explanations by Wu Liang, the garden's owner. While looking at Zhang Hong's representations, we can certainly imagine the delicate sweet fragrance of all those blossoms wafting through the Garden of Repose, inviting visitors to relax.

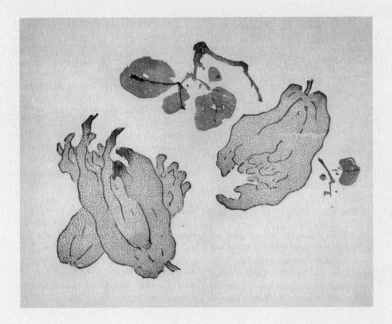

And we're not out of places to search for traces of Chinese citrus
culture. The collection of work from the Ten Bamboo Studio of Hu
Zhengyan (ca. 1584–1674), which dates from the late Ming dynasty,
includes a series of woodcuts of fruit, including the remarkable
variety of citron known as Buddha's hand or Buddha's fingers (*Citrus medica* var. *Sarcodactylus*). This fruit grows in a cluster of highly
irregular finger-like shapes and has long been popular in the Far
East, especially because of its pleasant fragrance.

What other clues to the origins of citrus fruit can we find? In
his memoirs, the Mughal emperor Jahangir, a seventeenth-century
ruler of India, waxed extremely enthusiastic about a shipment of
oranges (*kauṇlā*) from Bengal:

And though that place is 1,000 kos distant most of them arrived quite fresh. As this is a very delicate and pleasant fruit, runners bring by post as much as is necessary for private consumption, and pass it from hand to hand. My tongue fails me in giving thanks to Allah for this.

To date, no oranges have ever been found growing in the wild. One hypothesis is that they resulted from the cross-breeding of mandarins and pomelos. In any case, many people have sweated over the puzzle of the orange's origins. The experience of the botanist Emanuel Bonavia can illustrate some of the difficulties they faced. Bonavia visited the orchards of Delhi to learn more about the so-called suntara orange. But when he arrived, he was disappointed by what he found: the orange trees that made up the "orchard" were actually scattered among the trees of the forest. The fruit farmers justified this setup by claiming that the orange trees liked shade and grew especially well under their taller counterparts. Undeterred, Bonavia continued his research. In his book *The Cultivated Oranges and Lemons etc. of India and Ceylon* (1888) he wrote, "I have left no stone unturned in order to get at the bottom of the origin of this sùntara orange, and possibly at the derivation of its name."

Although no one could provide Bonavia with a final answer to the question of where the fruit came from, a certain consensus did exist on the topic. He reported, "All agree that it is not indigenous, and the legend has it that Hanuman, a general of Rama, introduced the plant on his return from Lanka (Ceylon). Some people say seeds were brought from Assam proper." At a later point, however, he offered a different origin story when he noted that most authors believed oranges originally came from China or the region known at that time as "Cochinchina," which today encompasses south Vietnam and eastern Cambodia. One thing is certain: oranges have been widespread in South and Southeast Asia for a very long time.

Today, nearly a century and a half later, we have a few more pieces of this puzzle. Studies by U.S. and Spanish scientists have narrowed down the list of potential birthplaces of citrus fruit. To do so, the researchers analyzed the genomes of more than fifty species, from the Chinese mandarin to the sour Seville orange, and even prepared a citrus family tree. The most important species in it are the citron, mandarin, and pomelo. And the place of origin has been narrowed down to the eastern and southeastern foothills of the Himalayas, a region of mild winters, intense sunlight, and relatively little rain. Tropical environments with high humidity make citrus fruit more susceptible to a host of diseases, so any place farther south can be ruled out, and most citrus plants do not tolerate frost.

LIKE THEIR COUNTERPARTS IN ancient China, Italian artists and collectors there were drawn to the sometimes irregular, fantastic forms citrus fruit takes. The powerful Medici family were particular citrus fans. In January 1665, the scientist and poet Francesco Redi sent a report to Cardinal Leopold de' Medici:

*As I intended to make some observations of the citrus fruits in
Your Honor's enchanting grove today I found a new, completely
novel variety and I asked one of the gardeners to cut off a few of
the bizarre fruits and send these novelties to you.*

Cosimo III, the Grand Duke of Tuscany and a member of the
Medici dynasty, had a similar citrus collection in Castello, near
Florence. He hired the Florentine artist Bartolomeo Bimbi (1648–
1729) to paint four different views of the resulting bounty. The
images are bursting with oranges, citrons, lemons, bergamots,
and limes hanging from branches or bushes.

As early as the sixteenth century, the rich and famous of central
Europe were swept up in the citrus cultivation craze. Just as in Italy,
a citrus garden was a sign of prestige. One such garden was located
in Augsburg—in southern Germany, not far from Munich—and
belonged to the Fuggers, an unfathomably wealthy commercial
dynasty. In 1531, this garden was said to contain examples of every
type of citrus fruit grown in Italy.

The practice of setting up temporary shelters for the trees,
which was common around Lake Garda, spread to royal courts
and grand houses farther north. These simple wooden structures
were insulated with moss and dung and heated with ovens, and in
the spring they could easily be taken down. Permanent buildings
designed to be used as orangeries did not exist until the end of the
seventeenth century.

While citrus fruit captured countless imaginations, Johann
Christoph Volkamer (1644–1720) earned a reputation for being
especially obsessed. Volkamer was born into a family with medical and botanical credentials. The title of his encyclopedic 1708
book on citrus fruit, *Nuremberg Hesperides*, refers to an episode in
the saga of Hercules: the hero steals the legendary golden apples

right
Citrus fruits domi-
nate the imaginary
landscapes painted
by German botanist
Johann Christoph
Volkamer (1644–1720).

Limon della Costa grosso. 369

of the gods guarded by the Hesperides, the three nymphs Aegle, Arethusa, and Hesperthusa. To his enchanted readers, the fruit featured in the book's pages must have indeed seemed to be the golden apples of Greek mythology. And "Hesperides" became a catchword for the rich garden culture that developed in the area around Nuremberg. Volkamer's own garden was one of the most magnificent in southern Germany.

The more than one hundred copperplate illustrations in *Nuremberg Hesperides* are unique in that they depict oversized citrons, lemons, bergamots, limes, and regular and bitter oranges floating in front of or above garden panoramas or views of buildings. In his garden in Gostenhof, near Nuremberg, Volkamer designed and installed a building for his fruit trees that was quite advanced for its time: it was open on its the southern end and the roof could be removed in the warm months. This structure was joined to a "fragrance chamber" where visitors could enjoy the sight and smell of blossoms and fruit during the winter. The illusion of a Mediterranean garden in northern climes was nearly perfect. Volkamer was known for his enterprising nature in Italy as well, but his reputation there had nothing to do with citrus groves. Instead, he owned mulberry orchards with silkworms and a neighboring silk factory.

below
Lemon harvest at Lake Garda, Italy, early twentieth century.

The dream of an idyll among the citrus trees captured imaginations farther north as well. In 1730, the Scottish poet James Thomson wrote the following paean to the delights of a fruit grove in his work *The Seasons*:

> *Bear me, Pomona, to thy citron grove,*
> *To where the lemon and the piercing lime,*
> *With the deep orange, glowing through the green,*
> *Their lighter glories blend. Lay me reclined*
> *Beneath the spreading tamarind, that shakes,*
> *Fanned by the breeze, its fever-cooling fruit.*

After describing the delights of citrus in Weimar, Goethe encountered his first sweet-smelling lemon groves around Lake Garda, a region with a citrus tradition stretching back into the thirteenth century. His travel book *Italian Journey* recounts an experience from September 1786:

> *We passed Limone, whose terraced hillside gardens were planted with lemon trees, which made them look at once neat and lush. Each garden consists of rows of square white pillars, set some distance apart and mounting the hills in steps. Stout poles are laid across these pillars to give protection during the winter to the trees which have been planted between them.*

Two months later he was near Rome and despite the winter season the trees were still delighting him:

> *Here you do not notice the winter. The only snow you can see is on the mountains far away to the north. The lemon trees are planted along the garden walls. By and by they will be covered*

with rush mats, but the orange trees are left in the open. They are never trimmed or planted in a bucket as in our country, but stand free and easy in the earth, in a row with their brothers. You can imagine nothing jollier than the sight of such a tree. For a few pennies you can eat as many oranges as you like. They taste very good now, but in March they taste even better.

Farther south the Sorrento Peninsula juts into the Gulf of Naples, not far from Capri and the antique ruins of Pompeii. Being there is like stepping into a fully realized cliché of the romantic Italian south. It's a picturesque landscape full of lemon and orange groves. Clinging to steep inclines, surrounded by ancient stone walls, growing on sunny terraces, and sometimes propped up on supports made of chestnut wood, the trees bloom practically the whole year round.

Among the many visitors these trees have inspired over the decades or even centuries was the philosopher Friedrich Nietzsche (1844–1900). His thoughts took flight as he wandered among them, enjoying their shade and glimpsing the white blossoms through their dark leaves. From October 26, 1876, to early May 1877, Nietzsche, then thirty-three years old, was a guest at the Villa Rubinacci.

Nietzsche came at the invitation of the writer Malwida von Meysenbug, who suggested that he spend the winter there with her. He had been relieved of his duties as a professor at the University of Basel for a year on account of his intense migraines and urgently needed to recover. Accompanying him on his journey were Paul Rée, his physician, and Albert Brenner, one of his students. From his room on the ground floor, the young philosopher could look directly into a garden. In a letter to her daughter, Malwida described how the olives and orange trees in the garden grew together to give the impression of a forest. When Nietzsche ventured outside, the trees sheltered him from the bright sunlight and the painful headaches it triggered. One especially interesting detail recorded about his stay is the particular importance that one of the trees developed for him. Ideas supposedly occurred to him whenever he stood under it, and it came to be known as the "thought-tree."

During his months at the villa Nietzsche worked on his seminal work *Human, All Too Human*. While the book makes no specific mention of the lemon, orange, and olive trees that surrounded the author while he wrote it, it does contain the observation "We like to be out in nature so much, because it has no opinion about us." What else might Nietzsche have experienced near Sorrento? Did he taste limoncello, the sugary sweet yellow lemon liquor that is one of the region's specialties?

It can't be a coincidence that so many writers have been inspired by lemon and orange groves. The blossoms' bewitching fragrance surely plays a role in their appeal. One testament to their power comes from the French author Guy de Maupassant (1850–1893), who traveled to the South of France to breathe in the scent of orange blossoms above the city of Monaco:

> Have you ever slept, my friend, in a grove of orange trees in flower? The air that one inhales with delight is a quintessence of perfumes. The strong yet sweet odor, delicious as some dainty, seems to blend with our being, to saturate us, to intoxicate us, to enervate us, to plunge us into a sleepy, dreamy torpor. As though it were an opium prepared by the hands of fairies and not by those of druggists.

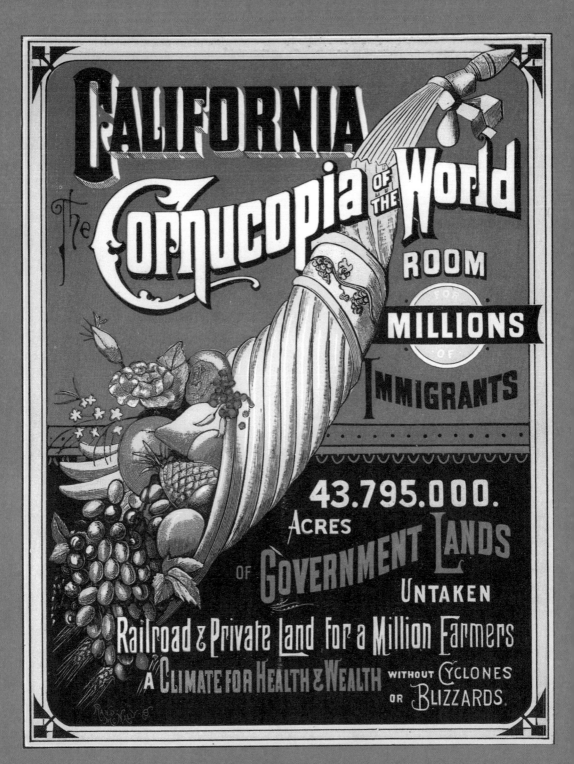

— 12 —

As American as Apple Pie

THE DEVELOPMENT OF ORCHARDS in North America followed a path that was different in many respects from the one taken in Europe. To start with, wild fruit was plentiful in the continent's vast forests until much closer to the present day. Wild fruit, berries, and nuts formed a central element in the diet of many of the First Nations. One indirect account can be found in *New Voyages to North America* (1703), by Louis Armand, Baron de Lahontan, a Frenchman who traveled though parts of present-day Canada and the United States. It describes how the Huron tribe in what is now Upper Michigan picked wild grapes, plums, cherries, cranberries, strawberries, blackberries, raspberries, and blueberries, and even preserved crabapples in maple syrup.

The British brought their farming techniques with them and the areas that they first settled on the East Coast had a highly developed fruit culture even in the colonial period. In 1629, Captain John Smith mentioned Jamestown's apple, peach, apricot, and fig trees. William Berkeley, the first governor of Virginia, supposedly raised about fifteen hundred fruit trees on his homestead, Green Spring. And in a 1656 book with the remarkable title *Leah and Rachel; or, The Two Fruitfull Sisters, Virginia and Mary-Land*, John Hammond, a resident of Virginia, wrote:

> *The Country is full of gallant Orchards, and the fruit [is] gener-*
> *ally more luscious and delightful than here, witnesse the Peach*
> *and Quince. The latter may be eaten raw savourily, the former*
> *differs and as much exceeds ours as the best relished apple we*
> *have doth the crabb, and of both most excellent and comfortable*
> *drinks are made. Grapes in infinite manners grow wilde, so do*
> *Walnuts, Smalnuts, Chesnuts and abundance of excellent fruits,*
> *Plums and Berries not growing or known in England.*

Another source of information is Hector St. John de Crèvecoeur (1735–1813), who was born in Normandy and worked for several years as a mapmaker in Canada, where he also fought against the British. Later he Anglicized his name to John Hector St. John and settled with his wife in Orange County, New York, an area with large numbers of fruit trees. The detailed descriptions in his book *Sketches of Eighteenth Century America: More Letters From an American Farmer* provide a fascinating glimpse into the apple business of the time and are well worth quoting at length. At one point he mentions "having planted in the fall a new apple orchard of five acres consisting of three hundred and fifty-eight trees." What would be done with all that fruit? "It is not for cider, God knows!"

de Crèvecoeur insisted. Presumably he wanted to head off the assumption that he was looking to drink all the time (although the temperance movement was still a long way off). In fact, the apples were primarily intended for pigs:

> *As soon as our hogs have done with the peaches, we turn them into our orchards. The apples as well as the preceding fruit greatly improve them. It is astonishing to see their dexterity in rubbing themselves against the youngest trees in order to shake them. They will often stand erect and take hold of the limbs of the trees in order to procure their food in greater abundance.*

After the harvest there was much work to be done. Maintaining good relations with the neighbors was therefore essential, especially when it came to preparing the fruit for drying:

above
A cider and apple stand on Lee Highway in what is now Shenandoah National Park, Virginia, 1935.

The neighbouring women are invited to spend the evening at our house. A basket of apples is given to each of them, which they peel, quarter, and core. These peelings and cores are put into another basket; and when the intended quantity is thus done, tea, a good supper, and the best things we have are served up. Convivial merriment, cheerfulness, and song never fail to enliven these evenings; and though our bowls contain neither the delicate punch of the West Indies nor the rich wines of Europe, nevertheless our cider affords us that simpler degree of exhilaration with which we are satisfied.

On the next day, the work continued as the group set up a wooden construction for drying the apples:

When the scaffold is thus erected, the apples are thinly spread over it. They are soon covered with all the bees and wasps and sucking flies of the neighbourhood. This accelerates the operation of drying. Now and then they are turned. At night they are covered with blankets. If it is likely to rain, they are gathered and brought into the house. This is repeated until they are perfectly dried.

De Crèvecoeur further explains how the fruit is used later on; it is placed in warm water overnight so that it swells back to its original size:

> When cooked either in pies or dumplings, it is difficult to dis-
> cover by the taste whether they are fresh or not. I think that our
> farms produce nothing more palatable. My wife's and my supper
> half of the year consists of apple-pie and milk. The dried peaches
> and plums, as being more delicate, are kept for holidays, frolics,
> and such other civil festivals as are common among us.

He also mentions a method for distilling apple wine into a highly concentrated liquor. Another specialty was apple butter, "in the winter a most excellent food, particularly where there are many children":

> For that purpose, the best, the richest of our apples are peeled
> and boiled; a considerable quantity of sweet cider is mixed with
> it; and the whole is greatly reduced by evaporation. A due pro-
> portion of quinces and orange peels is added. This is afterwards
> preserved in earthen jars, and in our long winters is a very great
> delicacy and highly esteemed by some people. It saves sugar and
> answers in the hands of an economical wife more purposes than
> I can well describe. Thus our industry has taught us to convert
> what Nature has given us into such food as is fit for people in
> our station.

Although the traces of most eighteenth- and nineteenth-century gardens have disappeared, period advertisements can give us a sense of how popular fruit trees were during these periods. Nota-ble examples come from a grower named William Prince, who

As American as Apple Pie

operated America's first full-fledged commercial nursery on eighty acres (thirty-two hectares) of Long Island. A 1771 newspaper contains a two-page spread listing the 180 types of fruit trees and plants he had for sale, mostly imported from Europe. The Old American Nursery, as the business was called, existed until the second half of the nineteenth century and published increasingly extensive catalogs throughout that time. The 1841 edition, for example, featured 1,250 types of fruit. In addition to running the nursery, the industrious Prince family published a number of related books, such as *The Pomological Manual; Or, a Treatise on Fruits: Containing Descriptions of a Great Number of the Most Valuable Varieties for the Orchard and Garden* (1831).

Even U.S. presidents were among the fans of fruit farming. In 1760, George Washington planted thousands of fruit saplings at his Mount Vernon estate in Virginia. Many passages in his diaries record his fruit farming activities, such as the grafting of diverse types of trees. The harvest was eaten directly, preserved, or processed into cider. One of the earliest entries on this topic comes from March 24, 1762, and provides a good sense of how much detail he recorded:

—
left
American Homestead Autumn by Currier and Ives, 1868/69.

Grafted 5 others of the same Cherry's on Scions standing in a
Cluster in the Mint bed.

Also, 3 Bullock hearts (from Colo. Mason) one under the Wall to
the right of the gate—2 others under the Wall also, between the
5 Cornation Cherrys & opposite to the Plumb trees. [Bullock's
Heart is the mildly sweet fruit of a tree called Annona reticulate,
which probably originated in the West Indies.]

In 1785, Washington rearranged the gardens at Mount Vernon. Fruit trees were relocated from the more formal upper garden, creating space for additional flowers and vegetables. He also obtained 215 apple trees from a Major Jenifer. Apples and other typical fruit trees for the estate, such pears, cherries, peaches, and apricots, were added to the other existing gardens and planted at the estate's surrounding farms. Some of these trees were trained as espaliers by the gardeners—who, we should remember, were slaves.

President Thomas Jefferson (1743–1826) shared Washington's enthusiasm for fruit farming. Large numbers of trees—representing 170 different fruit varieties—grew both at his famous Virginia estate Monticello and his other properties. The selection of apples was representative for the region at this time, including Hewe's Crab Apple, Esopus Spitzenburg, and Roxbury Russet. But his favorite was the type known as Taliaferro, which he claimed produced "the best cyder apple existing... nearer to the silky Champagne than any other." Sadly, we can only imagine the taste of this superior cider: the Taliaferro, like many historical fruit varieties, has been lost. In any case, Jefferson was a true American patriot when it came fruit, and asserted that Europeans "have no apple to compare with our Newton Pippin"—an earlier name for the celebrated Albemarle Pippin. Jefferson apparently let his enslaved

gardeners handle the work of grafting, and rather surprisingly believed that it was better to raise fruit trees from seeds.

Jefferson's enthusiasm could not change the fact that Virginia's warm, humid climate was too much for many European fruit species to handle. Pear, plum, almond, and apricot trees suffered from insect infestations and disease. But Jefferson still enjoyed a number of successes. He celebrated the Seckel pear, a variety from Pennsylvania, boasting that it "exceeded anything I have tasted since I left France, and equalled any pear I had seen there." The Marseilles fig earned similar praise: "the finest fig I've ever seen."

When the American Pomological Society was founded in 1848, one of the tasks it faced was to impose order on the sometimes messy business of selling fruit. Lower-quality varieties—either saplings or the fruit itself—were often offered under misleading names or established ones were simply rebaptized and promoted as something "new." The market was full of horticultural counterfeits that shed a bad light on the entire industry. In response, a

above
Boxes of peaches leaving the orchard, Delta County, Colorado, 1940.

committee of the society began working on a catalog listing the various names and synonyms for the different fruit types.

Of course, it's impossible to capture all the nuances of a fruit's appearance with words alone: a reliable identification guide requires pictures. Charles M. Hovey (1810–1887), a fruit farmer in Massachusetts, engaged the artist William Sharp to prepare illustrations of apples, pears, plums, peaches, and cherries. The result was a two-volume work, *The Fruits of America* (1848–1856), which Hovey published in the society's name. Additional artists joined the project in the ensuing years. One noteworthy contributor was Joseph Prestele, who had been an illustrator at the Royal Botanical Garden in Munich before immigrating to the United States. By 1930, the last year of its publication, the work had featured about 7,700 watercolors by sixty-five different artists.

Julius Sterling Morton (1832–1902), who was the U.S. Secretary of Agriculture under President Grover Cleveland and governor of the territory that became Nebraska, was not only a political figure but a passionate gardener. When he moved with his wife to Nebraska City, he bought a 160-acre (65-hectare) farm and began experimenting to see what would best grow under the local conditions. According to one of their contemporaries, the Mortons had two thousand fruit trees with the best apple, peach, and plum varieties available. In 1872 Morton issued a stirring call to make fruit farming a key element of the country's westward expansion:

> There is comfort in a good orchard, in that it makes the new home more like "the old home in the East."...Orchards are missionaries of culture and refinement. They make the people among whom they grow a better and more thoughtful people. If every farmer in Nebraska will plant out and cultivate an orchard and a flower garden, together with a few forest trees, this will

become mentally and morally the best agricultural State, the grandest community of producers in the American Union… If I had the power I would compel every man in the State who had a home of his own, to plant out and cultivate fruit trees.

On the very day that Morton gave this speech a meeting of great importance for trees just happened to take place as well. It concluded with the decision to designate April 10 as Arbor Day. This holiday is now an established tradition, although the precise date varies from region to region depending on the growing period. For the Great Plains, including Nebraska, fruit trees were even more important than elsewhere because there were practically no forests in the region. Before white settlers arrived in Nebraska, trees covered just 3 percent of the land, mostly in areas along rivers. By 1890 they had planted several million trees there, completely changing the character of the landscape in many places. On the farmsteads, trees were especially useful in providing protection from strong winds.

This enthusiasm for fruit trees is also notable because a significant break had occurred in the middle of nineteenth century. As the temperance movement picked up steam, more and more farmers neglected their apple orchards. But the drop in demand for hard cider was not purely due to people giving up alcohol. As the number of German immigrants increased, so did the brewing of beer—a beverage that apparently hit the spot with more Americans.

And yet—while it's not clear if Morton was solely responsible—American apple cultivation came roaring back. At the beginning of the twentieth century, the U.S. Department of Agriculture listed more than seventeen thousand apple varieties in its *Nomenclature of the Apple*. A decisive change made this turn of events possible: the fruit was increasingly seen as a healthy snack, and the types of apples earlier used to make hard cider and applejack, or apple brandy, barely played a role. This shift in the sorts of apples planted continued after the temperance movement reached its peak in 1920 with Prohibition. From that point on, most apple farmers produced nonalcoholic "sweet cider"—essentially unfiltered apple juice.

Ralph Waldo Emerson surely knew that cultivated apples were not native to America, but this did not stop him from claiming that "the apple is our national fruit...Man would be more solitary, less friended, less supported...[without] this ornamental and social fruit." And apple pie? While she is better known as the author of *Uncle Tom's Cabin*, Harriet Beecher Stowe also had thoughts about this ultimate American treat. In 1869 she wrote that "the pie is an English tradition, which, planted on American

right
Washington's lush apple orchards have long provided a steady supply of American apples. When Cragg D. Gilbert developed this label for his orchard in the early 1950s, he included Mount Shuksan, the mountain he most enjoyed climbing in Washington's North Cascades.

soil, forthwith ran rampant and burst forth into an untold variety of genera and species." Apple pie has existed in America since the sixteenth century, but the saying "as American as apple pie" was first recorded around 1860.

When it comes to the spread of apples in North America, one figure clearly stands head and shoulders above the rest: John Chapman (1774–1847) from Longmeadow, Massachusetts. In 1797, at the age of twenty-three, he supposedly set out with a seed-filled backpack. According to legend, Chapman—or, as he came to be known, Johnny Appleseed—always went barefoot, dressed in a coffee sack, and slept in hollow trees. His ever-westward journey took him beyond the frontier, where he would plant seeds, sell the settlers the young trees after two to three years, and move on. His endeavors were aided by the fact that land companies such as the Ohio Company of Associates made settlers' permanent claims to their land contingent on them planting at least fifty apple and twenty peach trees within three years.

above
Before the advent of freezers, fruit was often canned. A glimpse into Mrs. Botner's cellar, Nyssa Heights, Malheur County, Oregon, 1939.

When not busy with his apple business, Chapman regularly read aloud from the work of Emanuel Swedenborg, a Swedish theologian and mystic. Swedenborg not only claimed to have conversed with angels and spirits but also believed that all material things represent elements of the spiritual world and therefore should not be manipulated. As a result, Chapman—like Jefferson—disapproved of grafting. Such a step was an act of interference in the great divine plan:

They can improve the apple in that way, but that is only a device of man, and it is wicked to cut up trees that way. The correct method is to select good seeds and plant them in good ground and God only can improve the apple.

However, we should remember that at this time, before Prohibition, apples were primarily fed to pigs or used to make hard cider, so no one expected them to be as tasty as today's eating apples.

A certain amount of skepticism about the myth of Johnny Appleseed seems justified. Was he really the martyr that many of the stories about him describe? Did he really raise all of his apple trees from seeds? And did the settlers that bought them really follow his instructions and do the same?

There are fewer uncertainties regarding a later development that again helped to transform an entire American region: the cultivation of oranges in southern California, between Los Angeles and Riverside. The first navel oranges were planted in the area in 1873, when saplings were brought from Brazil. Over the following decades the number of trees swelled to a million. Until the early 1880s, many vineyards existed alongside the orange groves, but they eventually fell victim to blight. Orange trees were susceptible to a disease of their own—an insect known as cottony cushion scale—but growers were able to get it under control with the help of the vedalia beetle, a type of ladybug imported from Australia. In 1886, Wolfskill Orchards sent a trainload of California oranges back east thanks to the Santa Fe Railroad, which had just expanded its network to Los Angeles. Both the state itself and its signature crop were marketed together with the slogan "Oranges for Health, California for Wealth." Soon the boxes used to transport the fruit were adorned with labels showing bright idyllic images.

Effective irrigation was a central challenge. It was essential to route the meltwater from the San Gabriel mountains through canals, avoiding disputes among the water's users before they could arise. The key figure in solving it was George Chaffey (1848–1932), a Canadian engineer from Kingston, Ontario. In 1862 Chaffey and his brother William bought land that would later be the site of the cities of Ontario and Upland. With the help of the Mutual Water Company, a cooperative, Chaffey provided the settlers with equal access to water from the San Antonio canyon, which was channeled to each property through cement pipes. This innovative system was just one of Chaffey's many contributions to southern California's infrastructure.

A variety of factors, including land prices that increased five times over during 1887 and a loss of control over sales channels, put California's citrus growers under enormous pressure. But after they pooled their strength as a cooperative at the start of the twentieth century, the industry was soon on the upswing again. By this time, the orange farmers understood that the trees need protection from the wind to thrive. Eucalyptus trees were often planted for this purpose: they grew so quickly that the sensitive orange trees could follow just two to three years later. The trees were positioned in rows with twenty-two feet of space (nearly seven meters) between them. Farm crews consisting primarily of Chinese, Filipino, Japanese, and Mexican workers protected the trunks with cornstalks to ward off jackrabbits. As soon as frost threatened the plants, large fires were set around the edges of the groves. The thick clouds of smoke were intended to keep icy air away from the trees.

The fruit and the boxes it was packed in—with their colorful labels showing scenes of sunny California—inspired more and more people to make their way to the fabled Pacific state, either as tourists or new residents. When the entertainment industry also found a foothold there, the fate of southern California was sealed. Land prices multiplied again and many fruit groves had to be abandoned. At that point orange farmers found a new home in the Central valley. Both California and Florida also became known for grapefruit. The rather unsuitable-seeming name of this citrus is presumably a throwback to a very old and long since vanished variety with fruit that indeed hung from the trees in grape-like clusters.

Of course, California is not the only fruit-growing hotspot on the West Coast: Washington State is famous for its apples, as is the neighboring province of British Columbia in Canada. The

below
Grapefruits are larger than oranges, but this large? Florida, 1909.

A Wagon Load of Grape Fruit, Florida.

beginnings of this vast industry were small: in 1847, the Quaker abolitionist Henderson Luelling (1809–1878) and his son-in-law William Meek set out from Iowa in a covered wagon full of young apple trees. They traveled to Oregon and California to plant apple orchards and then turned north to Washington. Later, when the railroad connection between the East and West Coasts was completed, Washington developed into the largest apple region on earth. The town of Wenatchee, surrounded by apple orchards, calls itself the Apple Capital of the World.

The ability to use the train to quickly bring the harvest to market was a boon to other types of orchardists as well. One fruit to benefit was the delicate peach, which has been cultivated in the United States since colonial times. In his 1997 novel *American Pastoral*, Philip Roth memorializes the role of these fruit-bearing trains when his main character recounts an (invented) railroad connecting New Jersey and New York City:

> A railroad line used to run up into Morristown from Whitehouse to carry the peaches from the orchards in Hunterdon County. Thirty miles of railroad line just to transport peaches. Among the well-to-do there was a peach craze then in the big cities and they'd ship them from Morristown into New York. The Peach Special. Wasn't that something? On a good day seventy cars of peaches hauled from the Hunterdon orchards. Two million peach trees down there before a blight carried them all away.

But we should also remember that the concentration of particular crops in specific regions is a relatively new phenomenon. Not all that long ago, anyone with a farm had an orchard. And before the population began to concentrate in the cities, nearly everyone had a farm. In New England, some of these traditional orchards still exist and can be visited today. Lyman Orchards, in Middlefield, Connecticut, is one example: its roots go back to 1741, when the main crop was peaches. Near the house of Jonathan Fisher, the first Congregationalist minister in Blue Hill, Maine (he held this job from 1794 to 1837), a heritage orchard has been laid out around a two-hundred-year-old pear tree apparently planted by Fisher himself. And the Lookout Farm, founded in 1650 by the Puritan missionary John Eliot and a small group of settlers in Natick, Massachusetts, is now a farm with sixty thousand trees. Eleven different apple varieties grow there, along with pears, peaches, and plums. Like many other orchards, the farm allows visitors to pick their own fruit. The "pick your own" approach was the brainchild of a farmer in New Hampshire in the 1920s. It caught on and today it is popular in many countries around the world.

There are other stories to tell. Italians immigrating to the United States in the latter part of the nineteenth century brought artichoke seeds, grapevine cuttings—and fig trees. The fig trees popped up in backyards even in places like Pittsburgh or Cleveland, where nobody would have expected them to take root, and people took great care to ensure their trees survived the winter. The much-cherished fruits helped people maintain an emotional connection with their homeland. This practice was so widespread that the presence of fig trees indicated places where Italians, soon to be Italian-Americans, were living. The surprising fact is that many of these trees survive to this day. Every year the growing fruits are protected with old onion bags to keep birds at bay. The Italian Gardens Project, a living archive of Italian-American gardens and their keepers, has been documenting the trees and their exact locations for more than a decade. This awareness helps Americans with Italian forebears keep their cultural heritage alive.

— 13 —

Orchards Unbound

IN THE TROPICS, THE forests are bursting with fleshy fruit, but there's a catch: the plants must be cared for and protected from disease and insects if people are to gain maximum benefit from this bounty. The domestication of trees probably followed a similar path to the one it took in more temperate climes. When people found fruit in the forest or jungle that they particularly liked, they dug up the tree it came from and planted it near their home. The chosen tree benefited from this relocation as well— it faced less competition and might even be fertilized with communal waste.

Many people in the industrial West harbor the romantic idea that tropical forests are untouched and should stay that way, but humans have lived with and changed these forests for hundreds if not thousands of years. Charles M. Peters, curator of botany at the New York Botanical Garden, has done fieldwork in tropical forests around the globe for decades. In *Managing the Wild*, he reminds us that "in many cases, human activity has actually increased the diversity in tropical forests by creating new habitats, selective weeding, and the introduction of new species."

But let's move back into the past. In Central America, agriculture and the settlements that came with it started about eight thousand years ago, later than it did in the Fertile Crescent. In the tropics of South America, another four thousand years passed before this turning point arrived. The first crop to be domesticated was corn. It grew in the river valleys—damp in the summer, dry in the winter—of western Mexico. A rich selection of fruit had existed since time immemorial, but ancient mangoes (first domesticated in India), bananas (whose first unequivocal evidence for cultivation is in Papua New Guinea), and other species were not perfect matches to their modern counterparts. They were much smaller, had more seeds, and most likely tasted somewhat different as well. Other fruit included pineapples, papayas, passion fruit (or maracujás), star apples (caimitos), guavas, peach palms, açaí, and cocoa—just to name the most common.

But while fruit is abundant in the tropics, the concept of an "orchard" quickly reaches its limits in parts of the world with different agricultural traditions than those of (or derived from) Europe and the Middle East.

WHEN THE CONQUISTADORES ARRIVED in Central America, they were surprised by the Aztecs' famous royal gardens, which were full of ornamental, aromatic, and medicinal plants tended by large numbers of workers. These gardens were often located on hills and mountains at sites near springs and caves. Modern researchers have brought much of the symbolic meaning of these

previous page
Jungle in South Asia
with breadfruit tree,
nineteenth century.

gardens to light, along with the myth-infused worldview of the Aztec people in general. Like many aspects of life, plants were associated with gods or even had a godlike function. One example is the *Quetzalmizquitl* (*Parkinsonia aculeata*), a spiny shrub with long leaves that resemble feathers. It was linked with Quetzalcoatl, the feathered serpent god who was one of the Aztecs' most important deities. Many of the plants in these gardens were used by shamans. Plants that primarily served as food could not be found here, however: the lower classes supplied such crops to the rulers as tribute. Francisco Cervantes de Salazar, a sixteenth-century Spanish clergyman and scholar, described "the gardens in which Montezuma went for recreation" in his *Crónica de la Nueva España*:

> *In this flower garden Montezuma did not allow any vegetables or fruit to be grown, saying that it was not kingly to cultivate plants for utility or profit in his pleasance. He said that vegetable gardens and orchards were for slaves and merchants. At the same time he owned such, but they were far away and he seldom visited them.*

But there are hints that gardens existed that more closely followed the pattern of mixing subsistence and ornamental plantings found in other parts of the world. The Spanish conquistador Hernán Cortés gives us a sense of what they were like in his reports to Emperor Charles v, collected in *Cartas de Relación* (translated into English as *Letters From Mexico*) (1519), which contain a description of the botanical garden of Huastepec (Oaxtepec):

> *It was the finest, most pleasant, and largest that ever was seen, having a circumference of two leagues, and a very pretty rivulet*

*with high banks ran along it from one end to the other... There
were lodgings, arbours and refreshing gardens and an infinite
number of different kinds of fruit trees; many herbs and scented
flowers. It certainly filled one with admiration to see the gran-
deur and exquisite beauty of this entire orchard.*

In Huastepec, gods representing fertility, reproduction, dance,
and song were honored.

For people who lived in the jungle, such as the Maya who
inhabited Central America since at least 2,600 BCE, cultivating
fruit trees around settlements could be as simple as finding fruit
they particularly liked, digging up the tree that produced it, and
planting it nearer to home. From the seeds of the breadnut or Maya
nut tree the Maya made flour for tortillas. The fruit of the allspice
tree seasoned their meals, and the gummy resin of the rubber tree
was formed into balls that they used in games. The kapok tree was
considered sacred and represented the organization of the world:
its roots reached down to the underworld while the gods sat in its
crown. These types of trees can still be found today around the
ruins of the ancient city of Tikal in northern Guatemala. Faced
with a seemingly endless drought, Tikal's residents abandoned
the site around 900 CE. While the definitions of "fruit" we have
used so far don't apply perfectly to all of these plant products, we
can still ask one of our recurring questions: Were these trees cultivated in gardens or did they grow wild?

THE AMAZON REGION, ESPECIALLY the rural areas, is home to more
than a dozen species of palms producing fruit that is still eaten
today. The U.S. naturalist Herbert Smith, who traveled through
Brazil in the 1870s, described them in memorable terms:

Straight up from the water the forest rises like a wall—dense, dark, impenetrable, a hundred feet of leafy splendor. And breaking out everywhere from the heaped up masses are the palm-trees by thousands. For here the palms hold court; nowhere else on the broad earth is their glory unveiled as we see it.

Alfred Russel Wallace, who was working on his own theory of evolution around the time of Darwin, even devoted an entire book to them—called, fittingly enough, *Palm Trees of the Amazon and Their*

Uses. Meticulously he supplied a drawing for each palm species, explained how the parts of the plants could be used, and described the color and taste of their fruits, not all of which are edible.

Early settlers may have reached this region twenty thousand years ago, when sea level was lower than it is today because an enormous volume of water was trapped in masses of ice. The water first reached its current position on Brazil's coast about five thousand years ago, so places near bodies of water—whether on the coast or along rivers—were certainly flooded over the course of time. When people abandoned the areas where they had been living, they left behind the nut and fruit trees they had planted near their homes. And in a sense these plantings continue to provide a kind of "fingerprint" of these earlier settlements. They exist in the form of *terra preta de índio* or Amazonian dark earth, which results from the use of vegetable carbon to infuse the otherwise relatively unfertile soil with nutrients. Scientists have identified the palm species açaí (*Euterpe olearacea*), tucumã-do-Pará (*Astrocaryum vulgare*), and the moriche palm (*Mauritia flexuosa*)—with fruit featuring an attractive diamond-shaped pattern—as reliable indicators of this type of earth. Other palms are known for their ability to withstand the fires of slash-and-burn agriculture.

"Many areas of the Amazon may appear 'pristine,' but they are actually old regrowth forests or mosaics of orchards within a forest matrix," explains the U.S. geographer Nigel Smith. Recent archaeological investigations have shown that as long as four thousand years ago people in the Amazon were burning bamboo forests and promoting the growth of trees such as palms, cedars, and Brazil nuts—all considered sustainable practices. Other studies suggest that indigenous peoples of the Amazon domesticated banana plants over the course of two thousand years.

The central Andes also provide evidence of many plants, trees, and fruit from pre-Columbian times, especially in the form of the ceramics produced by the Moche culture. Researchers are certain that the region's inhabitants cultivated pepper, corn, peanuts, potatoes, squash, and sweet potatoes. The presence of irrigation systems suggests that agriculture was highly developed, and tools such as hoes and digging sticks have been found. People may also have collected fruit in the wild, including avocados, guavas, papayas, and pineapples.

below
A plate of tropical fruit from a German encyclopedia, 1896.

Indigenous ways of life are under enormous pressure today. Native tribes have traditionally maintained a mosaic of small gardens or islands of trees along frequently traveled jungle paths. These patches are sources of food during travel and hunting trips. The idea that the indigenous groups living in the transitional areas between the jungle and the savannah were hunters and gatherers who did not tend to or cultivate crops was only dispelled in recent decades.

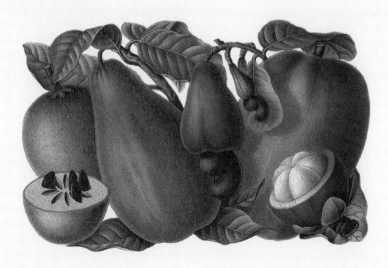

SEVENTEENTH-CENTURY BRAZIL WAS THE scene of an unusual horticultural experiment. On Antonio Vaz island, the Dutch field marshal and governor Johan Maurits (or John Maurice, Prince of Nassau-Siegen, 1604–1679) established a large tropical garden intended to provide a microcosm of the earth's natural history. It was located next to Mauritsstad, which was founded as an "ideal city" after the Dutch occupied the country in 1630. The garden's lay-out was also an attempt by the colonizers to impose geometric order on a region with flora that, to their eyes, was chaotically untamed.

Some of the plants originated in other parts of the globe, including the coconut palms that—planted to form an avenue—dominated the garden landscape. Coconut palms were first brought to northeastern Brazil in 1560 and quickly acclimated to their new home. The Dutch humanist Caspar Barlaeus described the monu-mental effort to transplant fully grown specimens to the garden:

> The Count ordered them to be fetched from a distance of 3 or 4 miles, in four-wheeled wagons, cleverly uprooting them and transporting them to the island, on pontoons set up across the rivers. The friendly soil accepted the new plants, transplanted not only with work, but also with ingenuity, and such fertility was passed to those aged trees, that, against the expectation of everyone, soon in the first year after transplanting, they, in a marvellous eagerness to produce, gave very copious quantities of fruit. They were already septuagenarians and octogenarians and because of this diminished the faith in the ancient proverb: "old trees are not for moving."

One section of the grounds was occupied by a quadratic orchard with citrus trees, while another featured pomegranate trees and grapevines. Many indigenous plants were included as well. A further highlight was a "garden of trees called banana." The Portuguese had recently brought banana plants from Guinea to South America, where they were known as "Indian fig trees." To this day, common small plantains are called "figs" in the West Indies. Maurits invited researchers from Europe to investigate the garden. Barlaeus recorded some additional impressions of what they might have encountered:

> *Beyond the coconut plantation, there was a place reserved for*
> *252 orange trees, in addition to the 600, which, grouped gra-*
> *ciously one beside the other, served as a fence and pleased the*
> *senses with the colour, the taste and the perfume of their fruits.*

Irrigation canals crisscrossed the site, which was also dotted with fishponds and enclosures for poultry. All sorts of animals were kept there, including wild boars and countless rabbits. It's hard to imagine that all these disparate elements added up to a harmonious whole, but apparently they did. If we believe our old friend Barlaeus, the garden was an earthly paradise—and one that maintained its remarkable attractiveness all year long:

> *The nature of the said trees is such that, during the entire year,*
> *they display flowers, mature fruit together with the green, as*
> *if one and the same tree were living, in various parts of itself,*
> *childhood, adolescence, and virility, at the same time.*

ON THE OTHER SIDE of the globe from Brazil lies Sri Lanka, formerly known as Ceylon. In this island nation, which was once nearly fully covered by rain forest, a fascinating example of a practical approach to fruit trees in the tropics has survived to this day. The most common type of agriculture there takes the form

of *gewattas*: gardens of miscellaneous fruit trees, herbs, and vegetables. Scientists first began looking more closely at their ecology in just the last few decades. One such gewatta expert is the Austrian biologist Karin Hochegger. When she published her study *Farming Like the Forest* in 1998, the gardens still covered more than an eighth of the island's land.

Dr. Hochegger's book takes a close look at 158 gardens. Within them she identified a total of 206 different species of trees. On average, each garden measured 56,500 square feet (5,250 square meters) and contained more than fifty different types of plants. A variety of birds, small mammals, and butterflies and other insects are at home there. Small paths wind their way among the plants. Typically, a well is located at a distance from the main building. Ferns—said to keep the water clean—grow on its walls.

Since no fences separate the gardens, it is difficult for a visitor to tell where one family's land ends and another's begins. Plants that bear valuable crops such as fruit trees and pepper vines (*Piper nigrum*), which climb up the small *Gliricidia sepium* trees, are usually located near the houses. As a result, they generally play less of a role in defining the property lines than other hedge-like bushes do.

Gewattas are especially typical for the region around the old royal city of Kandy in the island's interior, and they are generally located between three hundred and nine hundred feet (90 to 275 meters) above sea level. They left a lasting impression on Ernst Haeckel, a German evolutionary theorist who arrived on the island in November 1881 and spent four months there. He wrote:

> *The beautiful effect of the landscape found in the mid-elevation hills of Ceylon is that it takes on a role somewhere between garden and forest, between culture and nature. Sometimes one might have the impression of being inside a beautiful forest, surrounded by tall and splendid trees which are overgrown by various climbers. A small hut, however, partly covered by a breadfruit tree, or playing children, will remind us that we are inside a Ceylonese garden. The distinctive harmony between nature and culture is also revealed in the human component of these forest-like gardens.*

In earlier times, homes were not as spread over the landscape as they are today. People in the past had more reason to fear attacks from wild animals, and they lived closer together to protect themselves. The mix of trees in the gewattas also evolved significantly over time, influenced both by their owners' needs and the new species introduced by colonists and traders. For Hochegger, the gewattas have been in flux since their beginnings:

> *More than 2000 years of agricultural activities have led to a selection of many useful trees, shrubs and herbs which can be*

*grown close to human settlements. We can assume that the
earlier settlers and farmers collected fruits, nuts and resins in
the forest. In fact, most of the plants might have germinated on
their own after seeds were thrown away as kitchen waste. If a
new plant germinated near their hut they might have watched it
curiously eventually recognising what kind of tree was growing
there. Neighbours might have exchanged useful species, traders
brought new varieties and slowly the immense diversity of useful
species found in the gewatta today evolved out of trial and error.*

One typical gewatta plant is the jackfruit tree (*Artocarpus
heterophyllus*). Its fruit contains bean-like seeds that exude a pene-
trating scent when ripe but are edible. Because they contain a high
amount of starch, they can be used in place of rice. Jackfruit trees
require a good deal of space, and together with the tall-growing
coconut palms they form the "upper story" of the garden. Coconuts require twelve to fourteen months to ripen, but a fully grown
tree can produce fifty to eighty in a year. Below them are mango
trees, which originally came from the Indian subcontinent and
can be grown from seeds. Mangoes come in many different varieties: the fruit can be green, yellow, or red and take different
shapes. Orange and lemon trees are popular not only for their fruit
but because they are believed to keep insects away.

Gewattas also often feature guava, avocado, and breadfruit
trees; betel nut palms; and banana and papaya plants. The papaya,
which is actually a tree-like herb, was one of the species brought to
the island in the sixteenth century by the Portuguese, Dutch, and
English. The slightly oval aromatic fruit of the Bengal quince or
bael (*Aegle marmelos*) must ripen for about eleven months before
it can be split open with a hammer or machete. The slimy fruit,

which smells a bit like mango or banana, has a taste that reminds many people of marmalade or melted ice cream. It is often processed into a beverage.

The seeming chaos of trees, shrubs, vines, and herbaceous undergrowth is actually a symbiotic system in which each element has its place and fulfills a purpose. Cultivating these gardens, which flourish regardless of the season, is a matter of accepting what happens: "letting grow" or "what comes of itself" are the guiding principles. In other words, instead of trying to tame the wildness or combat certain plants designated as "weeds," the gardener leaves them be—motivated by the conviction that each part of the whole is there for a reason. And while the gewattas are left mostly untended, their yield is still surprisingly high, especially when we count not just food but additional products like firewood and seasonings. In many cases, the leaves, roots, seeds, or other parts of the plants found in gewattas are used in ayurvedic medicine as well. When I think about these gardens, I remember the words of my friend Abeyrathne Rathnayake, a sociology professor at Sri Lanka's University of Peradeniya. In his view, food is medicine.

You might expect that people would spend a lot of time in the gewattas, but it's not that common to run into other people when you are in one. They only come to gather fruit, herbs, or firewood. Sometimes they lay leaves or blossoms on the clean-swept paths to dry. Livestock is a rarity because most of the farmers are Buddhist and avoid eating meat or animal products.

John Davy, a British army doctor and the brother of the famous chemist Sir Humphry Davy, was hesitant to even apply the term "garden," let alone orchard, to the gewattas. In his 1821 book *An Account of the Interior of Ceylon and of Its Inhabitants*, he wrote:

Gardening among the Singalese is hardly known as an art; they plant, indeed, different kinds of palm-trees and fruit-trees round their houses, and flowering shrubs about their temples; and they occasionally cultivate a few vegetables, as yams, sweet potatoes and onions in their field; but in no part of the country is a garden according to our ideas to be seen.

From the outside, it's hard to guess what sort of thoughts occupy the people of Sri Lanka when they do spend time in their wild gardens. The writer Michael Ondaatje, who grew up on the

above
The Ceylon (now Sri Lanka) that German naturalist Ernst Haeckel saw and painted was (and is) the home of many plants unknown in the rest of the world, 1882.

island, reminisces in his autobiography about a mangosteen tree that he "practically lived in as a child." He also recalls a kitul tree by the family's kitchen:

> [The tree was] tall with tiny yellow berries which the polecat used to love. Once a week it would climb up and spend the morning eating the berries and come down drunk, would stagger over the lawn pulling up flowers or come into the house to up-end drawers of cutlery and serviettes.

The mixture of plants in these Southeast Asian gardens is very similar to what is found in the forests that surround them. As these forests disappear, the gardens increasingly provide a refuge for local species. It's tempting to view the gewatta as a throwback to "how things were done in the past." But there is another, more forward-thinking way of viewing them. If minimizing the environmental footprint of farming and fruit growing is a worthy goal in the twenty-first century, then it's hard to think of a more timely approach than the gewatta.

Just a four-hour flight east of Sri Lanka lies the Vietnamese island Thoi Son, in the Mekong Delta. Its abundant orchards form an incomparable fruit paradise—one that is traveled by boat. On Thoi Son fruit grows all year long in lush green orchards that have become an attraction for tourists: papayas, oranges, jackfruit, mangoes, durian, banana, pineapples, coconuts, and even apples and plums. Observers have reported that even the local fish have adapted to a fruit diet.

One curious detail of fruit farming in tropical regions comes from Java, where the small monkeys of the species *Macacus nemestrinus*, or pig-tailed macaque, were used to pick coconuts. In *A Naturalist* in *Borneo* (1916), Robert W. C. Shelford described this ingenious process, which is reminiscent of the fruit-picking monkeys in Egypt:

> A cord is fastened round the monkey's waist, and it is led to a coconut palm which it rapidly climbs, it then lays hold of a nut, and if the owner judges the nut to be ripe for plucking he shouts to the monkey, which then twists the nut round and round till the stalk is broken and lets it fall to the ground; if the monkey catches hold of an unripe nut, the owner tugs the cord and the monkey tries another. I have seen a Brok [the local name for a macaque] act as a very efficient fruit-picker, although the use of the cord was dispensed with altogether, the monkey being guided by the tones and inflections of his master's voice.

And since we're on the subject, let's consider coconuts for a moment and, by extension, the palms that they grow on. Coconut shells are extremely resistant to water—including saltwater—and floating coconuts can easily be carried by waves over the ocean. As a result, it's impossible to determine on which Pacific coast they were first farmed. Simply the question of whether they originated in the Americas or not was a hot topic among scientists for a century. Researchers now believe that they came to the New World with the Spanish (who first brought them to Puerto Rico) and the Portuguese (who brought them to Brazil), and that Southeast Asia was their original home. The demand for vast amounts of coconut oil to make soap—and the palm plantations, that is to say, "orchards" in the widest sense, to supply this demand—first arose about 150 years ago.

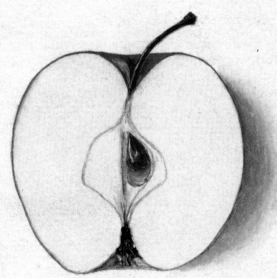

78455
Crab. (wild
J. Walter Basgye
Bowling Green.
Mo.

M. Strange.
11 - 10 - 14
12 - 15 - 14

— 14 —

Pomological Gentlemen

I N THE SIXTEENTH CENTURY, the pioneers of botany laid the foundation for encyclopedic efforts to record and research vegetation by beginning to categorize plants into families and species. Fortunately for us, they worked closely with artists of the age, who created countless portraits of these plants and their fruit. These images represent the zenith of botanical illustration, and their increasing precision and beautiful forms give a sense of the enthusiasm that plant classification inspired in many different countries.

Pomology—the science of growing fruit—crystallized as a discipline in the early nineteenth century and brought a whole new way of thinking about the different types of fruit. This new systematic study also had an interesting retroactive effect: although the term "pomologist" had just been coined, it suddenly seemed to apply to a range of people from the past. This long line of pomological forebears stretched from Theophrastus to Richard Harris (who founded

England's first commercial nursery in the sixteenth century) and
on to Thomas Andrew Knight (1759–1838, the president of the
London Horticultural Society) and beyond. Knight supported a
popular theory at the time that all fruit tree varieties have a pre-
defined lifespan and when its end point arrives they decay and die.
While this idea was quite wrong, it did inspire the cultivation of a
host of new cherries (including the Black Eagle, Elton, and Water-
loo varieties), apples, pears, plums, and other species according
to the most advanced scientific methods of the time to ensure a
supply of these fruits into the future. Knight's best-known book,
Pomona Herefordensis, was published in 1811.

From that point on, pomology was firmly established as a
subarea of botany. At the end of the eighteenth century and the
beginning of the nineteenth, most of the people attempting

to bring order to the sprawling diversity of fruit types were not full-time scientists, but rather clergymen, doctors, pharmacists, and teachers. They collected specimens, made drawings, and compared their findings. With lithography—which was soon even available in color—it became possible to duplicate images of fruit for relatively little money. One work to take advantage of this new technology was *Pomona Britannica; or, A Collection of the Most Esteemed Fruits at Present Cultivated in This Country*, which the draftsman and engraver George Brookshaw (1751–1823) first published in 1812. Brookshaw's book contains plates with impressive illustrations of the bounty produced by the British orchards of the day—256 varieties of 15 kinds of fruit. The images are so realistic that even today they make your mouth water. Many other authors produced important works on fruit in the decades that followed, including Robert Hogg (1818–1897), the editor of the London periodical *Journal of Horticulture*. His book *The Fruit Manual* (1860) was reprinted many times. During the same period the British Pomological Society was established with the goal of

> *promoting fruit culture in British dominions, especially to direct attention to the production of new varieties of fruits, examining and reporting on their merits and endeavouring to classify the fruits of Great Britain, the European Continent and America.*

Why was the world of fruit varieties such a tangled web in the first place? While fruit growers had known about grafting techniques since the Middle Ages, they didn't always follow the rules. Instead, when they needed new trees they sometimes just helped themselves to apple, pear, cherry, or plum saplings that had sprouted randomly. And if they liked the resulting plants, they used them as scions for the next round of grafts. If we imagine this process happening over and over at many different locations, we can imagine how a wide range of fruit types could emerge that varied from one place to another. Most of these local varieties never found their way into a pomological compendium.

France and Germany were pomological pioneers. In North America, in turn, the intensive study of fruit types didn't take off until the mid-nineteenth century. For a very long time, varieties were created more or less by chance. Targeted fruit breeding—in other words, the intentional application of the pollen of the father plant to the stigma of the mother plant—was not practiced until the twentieth century.

The study of fruit varieties branched out in sometimes fantastic ways. The clergyman and pomologist Johann Georg Conrad Oberdieck (1794–1880), who lived in Lower Saxony, claimed to have grafted together a tree that produced three hundred different kinds of apples. Oberdieck was apparently a man of extremes, as in addition to this botanical Frankenstein's monster he could also boast a collection of four thousand fruit trees of different kinds. An apple bears his name: Oberdieck's Rennet.

In the British Isles, the quirky result of urges, like Oberdieck's, to cultivate multiple types of fruit on a single plant is known as a "family tree." The person most associated with this pomological party trick is Leonard Mascall, who worked as a scribe for the Archbishop of Canterbury. Back in 1575, he provided some advice for those who wanted "to graffe many sortes of Apples on one Tree." Mascall assured his readers that "Ye may graffe on one Apple tree at once, many kinde of Apples, as on every braunch a contrary fruite, and of peares the lyke." At the same time, he issued a warning: "but see as nighe as ye can, that all your Cions be of lyke springing, for else the one wyll out growe and shaddowe the other."

right
Fig varieties
(White Hanover,
White Marseiles [*sic*],
Brown Naples or
Italian, Purple, Green
Ischia, and Bruns-
wick), 1812.

Carl Samuel Häusler (1787–1853), who went down in history as the first person in Germany to produce sparkling wine from apples, had very precise ideas about how to organize an orchard. These included a clear opinion about the placement of different kinds of fruit trees: they should never "alternate in close proximity or worse be completely mixed together." In his reasoning, trees "are like people: the family offers the most love and the place to thrive. For this reason, one should plant entire stretches with apple trees alone, others with just pears, and so on." It's not surprising that Häusler never provides a real explanation to back up his claim. But he did make his distaste known for the French practice of pruning the crowns of fruit trees into artificial shapes. Referring to the fact that wigs were all the rage in France, he laments:

> It is no wonder that the trees must wear wigs too, and we have grown so used to this form over the centuries that we now believe it must be so, and have stayed in the same rut until today, despite the fact that experience has shown us its disadvantages.

Another fertile field for pomological minds was identifying remedies for common tree ailments. In his book *The Orchardist, or, A System of Close Pruning and Medication, for Establishing the Science of Orcharding* (London, 1797), Thomas Skip Dyot Bucknall recommended a mix of corrosive sublimate (mercuric chloride) as an embalming agent and mold inhibitor, gin as a bactericide and solvent, and pitch as an antiseptic wound sealant. When the American Philosophical Society offered a prize of sixty dollars for a peach canker remedy, Bucknall immediately sent his book "across the Atlantic Ocean" and was content to "think myself happy, if the principles... improve the culture of American fruits." And William Salisbury offered some advice in his *Hints Addressed to*

Proprietors of Orchards, and to Growers of Fruit in General, published in London in 1816. It could be seen as an early example of recycling manufacturing waste:

> *Many useful manures may be obtained, such as
> the refuse of sugar-bakers, soap-makers, etc., etc.,
> bullock's blood, hair, and the scraping of seal skins,
> bone dust, and the refuse of the manufacturers of
> cart grease.*

ONE OF THE MOST remarkable fruit farmers of the twentieth century was surely Korbinian Aigner (1885–1966). This Bavarian priest faithfully served his ministry and congregation, but otherwise devoted all his attention to studying and planting fruit, especially apples. He considered fruit farming to be "the poetry of agriculture."

Aigner was a thorn in the side of the National Socialist authorities. A principled conservative who politically supported the anti-Nazi Bavarian People's Party, he often seized the opportunity to warn against the dangers of fascism, both from the pulpit of his church in the town of Sittenbach and elsewhere. He resisted an official order to fly the Nazi flag and made himself increasingly unpopular with those in power in a variety of ways. A dramatic turn finally came in 1939, when Aigner—who was a beloved leader of his congregation—made critical comments while teaching a religion class at a nearby school and was denounced by a teacher. He was arrested as a result and his odyssey through a series of prisons and concentration camps began.

In 1941 a group of clergy who had been critical of the government were transferred to the concentration camp in Dachau.

above
Apple breeder, Bavarian priest, and concentration camp survivor Korbinian Aigner.

Among them was Aigner, who was assigned to forced labor in an institute established there to research medicinal plants. During this period, he managed a seemingly unbelievable feat: he secretly bred apples from previously gathered seeds. Aigner gave these new varieties the names KZ-1 to KZ-4—all reflecting the place he found himself, as KZ is the abbreviation of *Konzentrationslager*, or concentration camp. He was most pleased with KZ-3, an apple suitable for eating directly or processing, and a helper smuggled bundles of Kz-3 saplings out of the camp. Later Kz-3 was rechristened "Korbinian." After the ss evacuated the camp, ten thousand prisoners were sent on a brutal march southward. The goal was supposedly an Alpine fortress in Austria's Ötz valley, but many of the marchers died on the way. Aigner was fortunate, however: he was able to escape and hide in a monastery, saving his life.

His passion for orchards and gardening continued to the end of his days and he earned a number of honors. He apparently rarely spoke about his time in the concentration camp.

We know so much about Aigner's work today because he made numerous color illustrations—drawn on everyday cardboard—of the many apples and pears that he studied. For him, these pictures simply served as a reference work to help him maintain an overview and ensure he correctly identified the different apple types.

A side view provides the first decisive data point. Is the fruit flat, spherical, hemispherical, a flattened sphere, round, conical, an oblong cone, a truncated cone, cylindrical, egg-shaped, bell-shaped, quince-shaped, or with the distinctive shape of a bergamot orange? The next step in identification is to consider the shape of the indentation at the base where the remnants of the blossom's sepals form the eye. Last but not least, the stem and the cavity around it provide valuable clues. But other factors influence what an individual fruit looks like, complicating matters. Fruit that grows on a standard-size tree appears somewhat different than fruit of the same type that grows on an espalier against a wall. Altitude and soil also each play a role. And of course even two apples or pears from the same tree are not perfect matches—and their appearance changes from week to week as they ripen.

After the Second World War, Aigner achieved near-legendary status as the "apple pastor." The illustrations that he created

above

KZ-3, also known as "Korbinian's apple," is one of the four apple varieties Aigner cultivated when he was held at the concentration camp Dachau during the Second World War.

constitute one of the world's most comprehensive collections of apple and pear illustrations—something Aigner himself may never even have realized. All in all, he left nearly one thousand such pictures to the institute for fruit farming in the Bavarian town of Weihenstephan, which is known for its wheat beer today. Experts still consider them valuable tools for fruit classification. And another account has been passed down to us from the end of Aigner's life: he was buried with the tattered coat that he wore as a prisoner in the concentration camp. The Korbinian apple, known for its fine balance of sweet and sour, is still grown today.

Despite successes like Aigner's, new apple varieties have always met with a bit of skepticism. Today you could be forgiven for thinking that some of them were bred to last long and look good rather than to taste particularly interesting. But there are also stories to discover, nuances of aroma, less-than-perfect shapes. We can harvest fruit from our own gardens at the point of perfect ripeness. Then it is most likely to be bursting with flavor, even when spots and wormholes disturb our dreams of flawless perfection.

When it came to fruit, Henry David Thoreau was skeptical of experts and scientists. "I have no faith in the selected lists of pomological gentlemen," he wrote. And not surprisingly, his judgment of the carefully cultivated fruits they favor is a harsh one: "They are eaten with comparatively little zest, and have no real tang nor smack to them." Conversely, the least elegant fruit could inspire him to flights of poetic fancy. Evoking the Roman goddess of fruit and orchards, he wrote, "Some gnarly apple which I pick up in the road reminds me by its fragrance of all the wealth of Pomona."

As a passionate hiker in every season of the year, Thoreau knew his way through the liminal realm of wild, feral, and cultivated fruit—"the old orchards of ungrafted apple-trees." He was intimately familiar with the gradations of their taste and smell, even when they were well past their prime. Deep in the winter he would come across forgotten fruit here or there that seemed worthy of consideration.

> *Let the frost come to freeze them first, solid as stones, and then the rain or a warm winter day to thaw them, and they will have borrowed a flavor from heaven through the medium of the air in which they hang.*

Thoreau could read all the signs that most others would ignore as insignificant. For him, a wild apple tree was like a wild child— and not any child but "perhaps, a prince in disguise." It is easy to see where his real sympathies lie when he declares, "It takes a savage or wild taste to appreciate a wild fruit." And he recommended eating such wild fruit right at the spot where it grows:

> *I frequently pluck wild apples of so rich and spicy a flavor that I wonder all orchardists do not get a scion from that tree, and I fail not to bring home my pockets full. But perchance, when I take one out of my desk and taste it in my chamber, I find it unexpectedly crude,—sour enough to set a squirrel's teeth on edge and make a jay scream.*

— 15 —

Orchards of the Senses

Come let us watch the sun go down
and walk in twilight through the orchard's green.

—RAINER MARIA RILKE, "The Apple Orchard"

MUCH LIKE FLOWERS, FRUITS come in an astounding range of colors and shapes that can move our imaginations in completely new ways. Giuseppe Arcimboldo (1526–1593), a Milanese painter, drew on these qualities in a unique fruit-based artistic experiment: not content to simply paint faces, he painted fruit and other parts of plants arranged into fantastical portraits. Opinion on the appeal of these "fruit faces" may be divided, but the idea behind them is certainly unique.

Earlier in this book I mentioned the magnificent gardens of China and the images of citrus fruit from the Ten Bamboo Studios. And indeed, the practice of painting fruit reached particular heights in China, developing into an art form of its own. The apricots, peaches, and plums that are so typical for the country can be hard for Western observers to tell apart—at least based on their descriptions—because some types of trees are related to both peaches and plums. But one English naturalist provides guidance on this topic: Thomas Muffet (1553–1604), otherwise known for his studies of insects, once said that "apricots are plums concealed beneath a peach's coat." When faced with the confusion these fruits present, who could argue?

In 1231, Sun Boren published his book *Portraits of the Plum-flower*. The tree in question is *Prunus mume*, also known as the Chinese plum or Japanese apricot. In one hundred ink drawings, Sun Boren captured the development of the tree's blossom from a bud to the moment when the last petal falls. Each drawing is accompanied by a poem expressing the mood of the piece. The black brushstrokes do not create a precise image: the point was to convey an essence of the plant, not record its every detail.

Although the act of painting fruit was valuable in itself as an exercise in contemplation, it most likely also led to a deeper examination of the emotional impact of the fruit's appearance on the artist. It was important to stop, observe closely, and capture the resulting associations on paper. If we turn our gaze inward to examine our own associations with orchards, what kind of art would the images that arise in our imaginations resemble? Surely something more emotionally charged than the precise illustrations of the pomological gentlemen. Maybe something along the lines suggested by Vita Sackville-West:

Some botanists consider the myrtle and the pomegranate to be
actually allied. Being no botanist I had merely remembered the
groves of myrtle and pomegranate in which I had slept in Persia.

Based on the many artworks that convey a vivid impression of
fruit trees and fruit groves, it would not be far-fetched to claim
that the period from the late nineteenth to early twentieth centuries was "the golden age of orchards." Most of these paintings
show scenes from the harvest season, but there are memorable
exceptions to this rule. For impressionists like Camille Pissarro or
Joaquín Sorolla, gardens and groves flooded with summer light
and the fragrance of blossoms offered a welcome contrast to the
demands of modernity: they were refuges, places for reflection,
and waking dreams full of sensory impressions.

below
In the Orchard,
1885/86, by Scottish
painter Sir James
Guthrie.

Few artists have taken the dreamlike aspect of traditional romantic orchards to such heights as the English artist Margaret Winifred Tarrant (1888–1959). Paging through the wonderful illustrations in her little books like *Orchard Fairies* and *Wild Fruit Fairies* has a magical effect, and practically convinces you that orchards are populated by countless sprites—half-human and half-butterfly—who know no greater pleasure than to cavort among the trees' heavily laden twigs and pluck the ripe fruit from them.

While Tarrant's illustrations lure us into an imaginary world, paintings of orchards can also be more representational, advancing agendas coursing through society. The 1893 Columbian Exposition—also known as the Chicago World's Fair—featured an outstanding example of one such work: Mary Cassatt painted a large mural for the main exposition hall of the Woman's Building. Unfortunately, it has been lost to the ravages of time with only an imperfect photo of it remaining. The center panel of this triptych showed a group of ten women and girls picking various kinds of fruit. It was called *Young Women Plucking the Fruits of Knowledge.*

The character and antiquity of an olive grove threatened by development beckoned to the French painter Pierre-Auguste Renoir (1841–1919). In 1907, he purchased the small estate on which it grew. "Les Collettes" (the name essentially means "an area with small hills") was located on the French Riviera, not far from Nice. The old farmhouse, with its weathered green shutters and a roof of terracotta shingles, was surrounded by picturesque vineyards, orange trees—and the olive grove. The 148 olive trees fanned out in a half-moon shape over a meadow laid out like a terrace so that the trees did not create too much shade. As a result, other plants had enough light and space to flourish. They included roses, carnations, and bougainvillea. And sweet-smelling lavender: the entire region was known for it, and it was gathered from

the fields for the famous perfume factories in Grasse, about twenty miles (thirty kilometers) away. Near the entrance to the house stood a strawberry tree (*Arbutus unedo*). While its fall-ripening fruit indeed resembles strawberries, it does not taste nearly as good. But the strawberry tree was not the only plant at Les Collettes that might inspire a double-take. The grounds also featured a tamarillo, known alternatively as a "tomato tree" due to its red, oblong fruit. Renoir's son Jean also remembered being amazed by a tree at the estate that produced both lemons and oranges.

right
Lovers in what appears to be an orchard, painted by Pierre-Auguste Renoir in 1875.

The Renoirs had greenhouses and cold frames that enabled them to grow flowers and vegetables year-round and sprout seedlings and saplings. Grapes and apricots were dried, while lemons, oranges, and tangerines were often left on the trees throughout the winter. Even the lemons developed a certain amount of sweetness as a result. Aline Charigot, Renoir's wife, had grown up on a vineyard between the Champagne and Burgundy regions and was in charge of the grapevines. Since they produced wine of middling quality, the family preferred to just eat the fresh grapes.

Renoir's workplace was a simple garden house nestled among the olive trees, with a corrugated metal roof and large windows in two directions. Cotton curtains helped him to optimally regulate the light that entered. The constant, calming hum of cicadas must have provided the soundtrack for his working hours. The view from the estate includes the medieval village Haut-de-Cagnes spilling down a steep incline. The whole place was a nearly inexhaustible source of inspiration.

He was apparently especially infatuated with the trunks of the olive trees, formed into irregular shapes by the countless storms and droughts they had withstood through the centuries. He had the dead branches removed, but otherwise let them grow wild and enjoyed the spectacle they offered. The trees had come to resemble a river of hardened lava, with hollows that provided an ideal spot for tiny-flowered sweet alyssum to put down roots. Painting these trunks was quite a challenge, even for Renoir:

The olive tree, what a brute. If you realized how much trouble it has caused me. A tree full of colours. Not great at all. How all those little leaves make me sweat! A gust of wind and my tree's tonality changes. The colour isn't on the leaves, but in the spaces between them.

The winter brought women and girls who spread out sheets below the trees and used long poles to knock the ripe black olives from the branches. Sometimes they also posed as models on the flower-dotted meadow, with the trees forming an indistinct background behind them. The harvested olives were brought to a press in the village and Renoir could not resist pouring oil from the first batch over warm toast and sprinkling it with salt. He was supposedly able to identify the oil from his estate just from the taste.

Occasionally Renoir picked out a single object such as an apple or an orange and captured it as a painted vignette. Other impressionist painters were equally taken with the pregnant qualities of fruit. Paul Cézanne once famously stated, "With an apple, I will astonish Paris." And Cézanne's paintings of apples indeed went down in history. His work *Rideau, Cruchon et Compotier* (1893/94) earned the title of the most expensive still life of all time after it was sold for more than sixty million dollars in 1999.

EMILY DICKINSON (1830–1886) was not just a gifted poet but, based on what we know of her life, an enthusiastic gardener with a solid knowledge of botany. The Homestead—the house on Triangle Street in Amherst, Massachusetts, where she lived with her family—still exists today, but subsequent owners made changes to the property. The flower and vegetable garden, the greenhouse, and the apple, pear, plum, and cherry trees all disappeared long ago. Archaeologists have spent years painstakingly excavating the soil layer by layer to determine how the garden looked in Dickinson's day, with the goal of recreating it in its earlier splendor. Meanwhile a small grove has been planted with traditional varieties of apple and pear trees: Baldwin, Westfield Seek-No-Further, and Winter Nelis.

The precise details of Dickinson's life are shrouded in mystery, but we know that at the age of thirty-eight she stopped attending church services. She penned some memorable lines on this subject that leave no doubt about the importance that the orchard had for her:

> *Some keep the Sabbath going to Church —*
> *I keep it, staying at Home —*
> *With a Bobolink for a Chorister —*
> *And an Orchard, for a Dome.*

The symbolism of orchards was not lost on other writers. The Russian playwright Anton Chekhov chose a country estate surrounded by a cherry orchard as the setting for one of his most famous tragicomedies—*The Cherry Orchard* (1904). While Chekhov's fictional orchard is magnificent, the trees no longer bear any fruit. The aristocratic owners, burdened by debt, think through their options: Why not cut down the trees and build cottages that could be rented to summer visitors? In the end, the orchard indeed meets this sad fate—not unlike the noble families, who no longer had a role in Russian society. The disappearance of the cherry trees stands for the loss of a way of life.

The rewards orchards offer those who own or toil in them are so much more than the financial gains the fruit provides. In chapter 6 we encountered William Lawson, the author of a popular seventeenth-century book on gardening. He admitted that the birds his fruit trees attracted—and the songs they sang—were almost as important to his enjoyment of his orchard as the fruit itself:

> One chief grace that adorns an Orchard, I cannot let slip: a brood of Nightingales, who with several notes and tunes, with a strong delightsome voice out of a weak body, will bear you company night and day. She loves (and lives in) hots of woods in her heart. She will help you to cleanse your trees of Caterpillars, and all noysome worms and flies. The gentle Robin-red-breast will help her, and in winter in the coldest storms will keep apart. Neither will the silly Wren be behind in Summer, with her distinct whistle (like a sweet Recorder) to chear your spirits.
>
> The Black-bird and Throstle (for I take it, the Thrush sings not, but devours) sing loudly in a May morning, and delights the ear much, and you need not want their company, if you have ripe Cherries or Berries, and would as gladly as the rest do you pleasure: but I had rather want their company than my fruit.

Later, the American poet John James Platt contrasted the ceaseless cacophony of nineteenth-century city life—"the roar of multitudinous wheels, the jarring of hammer, the trample of footsteps, all the sounds and voices of busy people ringing in our ears"—with the calm of the countryside with its "orchards ruddy and golden with their ripening fruit" and "cider mills swarmed about with half-tipsy bees."

Who wouldn't want to spend a warm summer afternoon in the shade of old fruit trees, perhaps with a book? (Maybe even this one?) For many people, an orchard evokes memories of the carefree days of childhood. Playing in the shadow of an apple tree, hardly able to wait for the fruit to ripen. Someone always gave in to the temptation to bite into one still-green, sour apple. As the summer drew to a close, the aroma of ripening fruit filled the air.

below
The Apple Gatherers
by English painter
Frederick Morgan,
1880.

Of the many people who have celebrated the orchard as a place of perfect peace, none has surpassed Virginia Woolf in the aptly named *In the Orchard*:

> *Miranda slept in the orchard, or was she asleep or was she not asleep? Her purple dress stretched between the two apple-trees. There were twenty-four apple-trees in the orchard, some slanting slightly, others growing straight with a rush up the trunk which spread wide into branches and formed into round red or yellow drops. Each apple-tree had sufficient space. The sky exactly fitted the leaves. When the breeze blew, the line of the boughs against the wall slanted slightly and then returned. A wagtail flew diagonally from one corner to another. Cautiously hopping, a thrush advanced towards a fallen apple; from the other wall a sparrow fluttered just above the grass. The uprush of the trees was tied down by these movements; the whole was compacted by the orchard walls.*

Orchards enchant whether heavy with fruit in the fall or bursting with blossom in the spring. Osbert Sitwell (1892–1962), the eldest of the legendary eccentric British poet Edith Sitwell's two younger brothers, recalled the details of a remarkable "garden party" in 1934 hosted by a prince in Beijing. The purpose of this gathering was to admire the crabapple trees in bloom. The conditions were perfect, as

> *the advance of the year was so rapid, you could almost hear the branches of apple and quince and wisteria creaking with the life within them, almost see the sticky buds first appear, and then unfold and open into their spice-breathing cups and tongues and turrets.*

When the elderly party guests arrived and alighted from their rickshaws, they found themselves in gardens that

seemed immense... Inside the boundaries of their walls, crowned with yellow tiles, were groves of old cypresses, the frond-like arrangements of their leaves lying upon the air as though they were layers of blue-green smoke, there were eighteenth-century water-gardens, now dry but full of wild flowers, and there were the sunk gardens wherein flourished, with gnarled, rough trunks, the crooked and ancient fruit trees which constituted the chief pride of their owner.

above
In the Orchard,
1891, by American
impressionist painter
Edmund Charles
Tarbell.

Far from simply taking in the view, the guests subjected the blossoms to intense scrutiny:

> *Slowly, painfully, the old men hobbled along the crooked, paved paths that zigzagged to these trees. When they reached them, they were conducted up small flights of stone steps, so fashioned that, saving where the steps showed, they seemed natural rocks that had cropped up through the turf or had fallen from the sky. These flights, their tops level with the tops of the trees, are thus placed near apple and pear and peach and quince and cherry, so that the connoisseur can obtain a perfect view of the blossoms. Even to a newcomer, inexpert in the flowery lore of the Chinese, from each different plane, the particular view of the tree for which the step had been constructed offered a revelation of a new world.*

Sitwell further related that after taking up their positions they spent an hour contemplating the blossoms and comparing their color and fragrance to those from the previous year. Only then did they gradually move on, enjoying the impression of the trees and the entire area they occupied.

Yet while orchards were calming, romantic places to spend time, their main purpose was to produce fruit. Making use of all that bounty was a lot of work and harvesting and processing fruit engaged the senses in a different way. The Swedish painter Carl Larsson (1853–1919) recalled the challenges that came with 1904's rich harvest: "From the middle of the summer we had to prop up the branches so that they did not break under the weight of the ripening Astrachan apples, Gravensteins, and whatever they were all called." The yield was so large that the family ate "applesauce, marmalade, and jelly… day and night for months"—to the point that they longed for bananas and dates.

above
Apple Harvest, 1888,
by Camille Pissarro.

MOST PEOPLE WHO TEND or occasionally visit an orchard are not "artists" in the conventional sense, but their often-intense connection to the trees and fruit may still express itself in ways that stand the test of time. Figures of speech, stories, superstitions—these are all artifacts left to us by generations of "everyday artists." It's sometimes impossible to determine whether the messages they convey ever had a basis in empirical reality, but even if they didn't, they exercised a power over people's imaginations and formed the basis of stories to be told and retold. Many examples come from central Europe, and while their precise origins have been lost, they are certainly several centuries old.

These cultural remnants point to a belief that fruit trees were conscious beings deserving of care and respect and that—in a continuation of ancient concepts—gardens were the abode of gods. People viewed the trees' development over the course of the year as running in parallel to their own lives and held that special practices could influence how they grew and how much fruit they bore. A few examples from the German-speaking parts of Europe provide a sense of this practical magic.

After feasting on Christmas Eve, it was advisable to bring the food scraps and nutshells to the fruit trees to "feed" them as well. Owners of apple trees had very specific duties to fulfill on January 6, Epiphany or the "twelfth day of Christmas": they needed to fill their mouths with fritters, kiss the tree, and say "Tree! Tree! I am giving you a kiss—now grow as full as my mouth." It was also common to wrap trees in straw on winter holidays such as Christmas Eve, New Year's Eve, or New Year's Day. Some researchers believe that this practice is the vestige of grain sacrifices performed in earlier times. Shaking, hitting, or knocking on the trees on certain holidays was also supposed to encourage them to produce more fruit. Could it really make a difference to let the trees "keep" every tenth fruit, wish them a happy new year, or hang a bone on a barren tree to shame it into being more productive?

When a young tree produced its first crop of fruit, the pickers should be sure to put the harvest in a big basket so that the tree could see what its owners expected in the coming years. Another way to ensure good harvests in the future was to give some of the fruit away, especially to a pregnant woman. At the same time, people in the Palatinate region believed that trees planted by a pregnant woman would not bear fruit.

Britain had some remarkable customs of its own, including "apple howling." In short, children would strike the trees with sticks and sing a variation of the traditional New Year's "wassail song." The point was to drive away evil spirits. Or was it? Folklorists suspect that the original purpose of the practice was to scare out harmful insects hibernating in the trees so that the birds would eat them.

Orchards are clearly treasure troves of tradition, practices, and stories. Should they really be banished to the edges of our property, as if they didn't deserve a place of honor at the center of domestic life? The opinionated William Cobbett expressed very passionate views on this subject in *The English Gardener* (1829):

> *I see no reason for placing the kitchen garden in some out-of-the-way place, at a distance from the mansion-house, as if it were a mere necessary evil, and unworthy of being viewed by the owner. In the time of fruiting, where shall we find anything much more beautiful to behold than a tree loaded with cherries, peaches, or apricots, but particularly the two latter? It is curious enough, that people decorate their chimney-pieces with imitations of these beautiful fruits, while they seem to think nothing at all of the originals hanging upon the tree, with all the elegant accompaniments of flourishing branches, buds, and leaves.*

Finally, what would a book about forgotten orchards be without a reference to Italy's "Orto dei Frutti Dimenticati," the "Garden of Forgotten Fruits" in Pennabilli? This magical place between Florence and San Marino once belonged to the Missionaries of the Precious Blood. It preserves Apennine fruit trees—apples, pears, quinces, cherries, and medlars—from disappearing from memory or even disappearing entirely. In addition, the garden features a number of sundials, dovecotes, and sculptures and other works by contemporary artists. There's also a very special highlight: the "Refuge of the Abandoned Madonnas," a collection of terracotta and ceramic statues of Mary. It's easy to imagine that these Madonnas, who once graced the votive shrines at crossroads in the countryside, have retreated to this peaceful spot to escape humanity's neglect and the transgressions of our age. A dreamlike fruit-filled sanctuary at the edge of the modern world.

— 16 —

Returning to the Wilder Ways of Fruit

T HE DAYS WHEN FRUIT growing primarily meant monks lovingly tending the trees in secluded monastery gardens are long gone. Most of the fruit sold in supermarkets today has never been anywhere near a romantic grove. Instead, it comes from vast plantations that operate more or less like factories, with the goal of producing fruit of consistent size and flavor that stays fresh as long as possible. In many cases, it can even be harvested before it is ripe. We should also remember that, for people outside the tropics, shortages of fresh fruit and vegetables during the winter months and into the spring were commonplace until the middle of the twentieth century. Increasing the availability of fresh fruit required long-distance transport and storage in controlled-nitrogen environments that keep it from spoiling. Now many of us have a vast range of choices. For example, although pomegranates don't ripen in the Northern Hemisphere until September, consumers there can buy them in the summer, too, because they are flown in from South America to be sold at local supermarkets. Of course, there are additional ecological costs to such a practice.

previous page
Handpicking is
still the best way
to harvest apples,
mid-twentieth
century.

Much of the fruit for sale today is not even fruitful in the literal sense: just think of all the seedless grapes and mandarins out there. Towering fruit trees with their majestic crowns are also disappearing. Modern farming methods and equipment are designed with more compact-growing varieties in mind, which are easier to tend and bring larger, higher-quality yields.

It's a development that could be described as "dwarfing down." If you walk through a modern apple farm, you will see long rows of spindly growths, each with just a single main branch. The apples themselves hang on the short, horizontal limbs that protrude from this central axis. This kind of high-intensity cultivation has nearly nothing in common with the traditional notion of a nineteenth- or twentieth-century orchard. At earlier points we looked at how gardeners in the past cultivated fruit trees to grow on walls. Today the term fruit wall stands for something quite different: a narrow hedgerow formed when trees are planted closely together and pruned mechanically. The point of this exercise is to save space and—thanks to a tractor that prunes the trees—labor so that the fruit can be brought to market cheaply.

Do methods like these represent the point of maximum fruit-farming efficiency, or are further improvements possible? Anthony Wijcik, of Kelowna in British Columbia, created a sensation in the mid-1960s with a surprising horticultural innovation.

His daughter Wendy had discovered a mutation in a fifty-year-old McIntosh apple tree: instead of branching into a progressively finer network of limbs and twigs, one of the branches produced fruit directly from short spurs. The "columnar" apple trees produced from this mutation are known as McIntosh Wijciks. Their apples are large and dark red but unfortunately not particularly tasty.

below
Apple harvests used to be a communal activity. Location unknown; the coloring is contemporary.

At the East Malling Research Station in Kent, England, the McIntosh Wijcik was bred with a hybrid of the English Cox's Orange Pippin and the French Court Pendu Plat to produce an ornamental apple tree known as Flamenco or Ballerina. It's small enough to keep on a balcony, but the apples are relatively susceptible to disease. The next step is the Minarette, which is not a specific variety but rather a style of pruning to keep trees narrow. Slender columnar trees are small enough that a good-sized balcony can accommodate an entire "orchard."

Other experiments involve hybrids of existing fruit species. The names sometimes take getting used to: *aprisali* is an apricot/plum mix, while a *peacotum* combines peach, apricot, and plum. *Pluots* are mostly plums, genetically speaking, but also contain other elements.

And is the Cosmic Crisp indeed "the most promising and important apple of the future," as the *New York Times* announced? In any case, this apple—which was first bred two decades ago at Washington State University—stakes its claim to this title based on its taste and long shelf life, and whole orchards in Washington are now devoted to its cultivation. Two-thirds of the entire U.S. apple harvest is grown in the state of Washington. Recent statistics show that just fifteen types of apples account for 90 percent of this fruit. The ubiquitous Red Delicious occupies the number-one spot.

In the past, the situation was quite different. An estimated seventeen thousand apple varieties were grown in the U.S. over the centuries, but thirteen thousand have disappeared, along with many of the family-owned orchards that could once be found in nearly every region of the country. "Those who have tasted them speak of them reverently with shining eyes, drawing in their breath with a gasp of keenly remembered delight," wrote the English author Philip Morton Shand in 1944. He was referring to the Somerset Pomeroy and the Court of Wick, apple varieties that have survived. Could it really be that every single one of the seventeen thousand types of apples had a distinctive taste? Maybe not, but surely a host of untold flavor sensations have been lost forever. Large fruit producers tend to concentrate on sweet apples, but appreciation for more complex-tasting varieties is on the rise again. The old English varieties Yarlington Mill and Kingston Black are just two examples of apples that offer an interesting mix of tart and sweet notes.

What do we lose when we breed fruit for mass consumption and aesthetic perfection? In *Epitaph for a Peach*, David Mas Masumoto vividly describes the pressure on fruit farmers to grow varieties that are especially popular at the moment. No matter how much orchard operators appreciate a certain type, they may have to switch to another, perhaps simply because consumers find a new lipstick-red color more appealing than the traditional yellow—or "amber gold" as Masumoto prefers to think of it:

> *The last of my Sun Crest peaches will be dug up. A bulldozer will be summoned to crawl into my field, rip each tree from the earth and toss it aside. The sounds of cracking limbs and splitting trunks will echo through the countryside. My orchard will topple easily, gobbled up by the power of the diesel engine and the fact that no one seems to want a peach variety with a wonderful taste . . .*
>
> *Sun Crest is one of the last remaining truly juicy peaches. When you wash that treasure under a stream of cooling water, your fingertips instinctively search for the gushy side of the fruit. Your mouth waters in anticipation. You lean over the sink to make sure you don't drip on yourself. Then you sink your teeth into the flesh, and the juice trickles down your cheeks and dangles on your chin. This is a real bite, a primal act, a magical sensory celebration announcing that summer has arrived.*

Stories like these show the importance of all efforts to preserve genetic variety. The British National Fruit Collection has had an eventful history, and at some points was in danger of being eliminated. But it continues to this day, and Brogdale Farm, where it is located, is home to a dizzying array of fruit varieties: 2,200 apples, 550 pears, 285 cherries, 337 plums, 19 quinces, and 4 medlars, plus 42 types of nuts (primarily hazelnuts). The collection's experts share their expertise with professional fruit growers. Fruit competitions in a number of places, especially the United Kingdom and the United States, promote lesser-known fruit varieties. And some research centers hold tastings that allow visitors to identify their personal favorites. For example, John Preece of the National Clonal Germplasm Repository in Davis, California, regularly offers an astonishingly wide selection of pomegranate and persimmon samples.

Not only are fruit varieties being saved, but a welcome resurgence of interest in more modest orchards is underway in many countries. Instead of focusing on squeezing the maximum profit from every bit of available land, these gardeners are motivated by the idea of something greater, more meaningful, and even more beautiful that can't simply be captured in monetary terms.

It's easy to fall in love with such smaller, more fathomable orchards. For the people who tend them, fruit is more than just a means to an end. Work on a large-scale fruit farm is honorable and important—after all, billions of people are alive today and they need to eat. But by definition these facilities lack the magic that people of earlier generations must have experienced when they worked with fruit.

And there seem to be few limits on the creativity of smaller fruit farmers today. In a garden in Ortakentyahşi, near the popular resort town of Bodrum on Turkey's Aegean coast, the owner Ali Cila explained to me that he can increase his harvest

above
Cracked-open
pomegranates set
aside as chicken feed,
Bodrum, Turkey.

by interspersing his pomegranate and olive trees. When he's so thoroughly convinced that this method works, who's going to contradict him? Cila also gets a kick out of growing two pomegranate varieties—one sweet and one sour—on the same tree. Fruit that splits open and can no longer be sold becomes a treat for the chickens, who stuff themselves with the seeds. Consciously or not, he's continuing a tradition that stretches back millennia: giving fruit that is no longer fit for human consumption to domestic or wild animals. Deer kept in a pen, for example, quickly smell apples when the wind is right and follow the appealing scent to its source.

Germany's Lower Franconia district in northwestern Bavaria—a region primarily known for its wine—is also home to a most unusual orchard, called Mustea. Marius Wittur, the owner, is dedicated to quince trees; specifically, traditionally cultivated varieties that are now in danger of being lost. When I first heard of this orchard, I was amazed that it was possible to find such a range of quince trees at all. But the fruit has become more and more popular in recent years. Perhaps people increasingly appreciate its not-so-sweet taste or are nostalgic for the time when the trees—along with medlars and mulberries—were much more common than they are today. In Portugal, this fascinating fruit is called *marmelo,* and it was there that I first encountered the quince paste known as *marmelada* that is cut into slices and eaten with cheese. And while most quinces are far too hard to eat raw, deep-yellow varieties grow in Turkey that you can bite into almost like an apple.

Efforts to recultivate this "lost fruit" are already something to celebrate, but the more I learned about the German project the more interesting it became. For one thing, the trees are grown according to the principles of organic farming. The idea is to concentrate on raising healthy trees rather than generating the largest possible harvest. This means that tree vitality is the goal—and as it increases so does the quality of the fruit. The farm does not even use artificial irrigation. Instead, care for the orchard takes the form of extensive tending of the landscape, based on the belief that the water naturally contained in the earth should determine the fruit's size and growth. This approach is seen as the only way to concentrate the quince's aroma in its authentic intensity.

Then the idea was added to incorporate grazing animals that eliminate the need to mow the grass under the trees. This gentle method of controlling the undergrowth has many ecological benefits. The Coburg Fox sheep, an old breed that has attracted new attention in recent years, seemed like a good choice: they used to be found in uplands with sparse vegetation that, from a climatological perspective, are similar to Franconia. The Coburg Fox especially stands apart from the rest of the sheep crowd when it is young: it starts life with wool that ranges in color from golden yellow to reddish brown. But it shares an advantage common to all of its kind: the storied "golden hoof." This poetic-sounding term refers to the fact that their hooves allow sheep to tread lightly over the land without significantly compacting the soil. Their tendency to nibble down blades of grass rather than ripping out whole clumps as horses do is also beneficial. The principles for growing the fruit are applied to the flock, as well: wool quality and meat yield receive less attention than vitality, hoof development, and behavior, and the farmers pay the most attention to these latter factors when deciding how to care for the animals.

Today plants and insects that have become rare sights elsewhere have taken up residence in the orchard's fields. And two praiseworthy endeavors exist in harmony. Quince varieties that were facing extinction are making a comeback, while an ancient breed of sheep thrives and multiplies in the rhythm of the agricultural year. Flora and fauna coexist in a new relationship that points the way to a better future for farming.

The quince trees cover twenty acres (eight hectares) and the approximately one hundred varieties—a "hodgepodge," according to the orchard's employees—are believed to make up the most extensive collection in all of Europe. Since it ripens at different times, the lemon-yellow fruit is harvested from the end of September to the last week of October. The fruit finds its way into a selection of wines as well as liqueur, juice, syrup, jelly, marmalade, and quince bread. And some of the lamb meat is turned into bratwurst with quince seasoning.

Adherents of approaches like this can be found in Great Britain as well, where the robust Shropshire sheep first used on German Christmas tree farms have been introduced to cider orchards. These animals have an excellent reputation as calm and docile. They tend to stick to the alleys between trees and, as long as their diet is balanced, they leave the bark alone. As good mowers, they keep the grass down while providing a new income stream. And by eating the fallen leaves in autumn, they may also reduce apple scab, which is caused by spores that can survive the winter on the ground beneath the trees. Other breeds may be suitable for

different contexts. The Olde English Babydoll, for instance, keeps the grass short in vineyards while leaving the grape leaves alone.

The use of sheep in modern fruit farming is just one example of the links between orchards and animals over the ages. At various points in this book we saw how animals large and small have played important roles in fruit tree cultivation or the spread of pits and seeds. Early on, wild horses and camels ate the fruit and eliminated the indigestible seeds or stones in far-flung locations. Monkeys were—and are—trained to climb high in the trees to pluck hard-to-reach fruit and deliver it to waiting human hands. Bees transport pollen to blossoms and their activities increase the fruit harvest. Farmers also introduce insects to fight pests, and chickens, ducks, and geese can play a part here too. But these birds, especially geese, can pose a danger to saplings. Those who tend orchards need to think carefully about what kinds of domestic animals they invite in. Dogs, incidentally, seem to offer an interesting benefit: scientists have discovered that they can sniff out huanglongbing, also known as citrus greening, a fatal bacterial disease spread by plant lice.

Today's orchardists continue to push the envelope while drawing on methods from the past. And sometimes they plant their orchards in the most unlikely places. One such orchard, the Krameterhof, lies in the heart of Austria about a hundred miles (160 kilometers) southeast of Salzburg. Instead of nestling in a protective valley, it perches on land located between 3,600 feet and nearly 5,000 feet (1,100 to 1,500 meters) above sea level—the kind of place where a barren landscape with scattered spruces is the norm. The textbooks tell us that fruit trees can't grow at altitudes of much more than 3,000 feet (1,000 meters). But this example proves them wrong. The surprised visitor can take in the sight of a Mirabelle plum tree bursting with ripe fruit, along with apple, pear, and other plum trees. The farmer, Sepp Holzer, even has kiwi trees—five different kinds. As unbelievable as it sounds, fourteen thousand fruit trees grow in the orchard. Look at that: there's an orange tree with a few ripe oranges actually hanging from its branches.

How does Holzer manage to grow fruit from southern climes here, despite the unfavorable climate in this high-altitude location? He starts by choosing a hollow protected by the wind as a tree location. Then he takes advantage of what he calls the "masonry stove effect" by placing large stones at the spot to store the sun's heat. "The stones 'sweat' and water condenses underneath them," explains Holzer, whose experimental approach has earned him the titles "agricultural pioneer" and "farming rebel." According to Holzer, "These moist spots are ideal for earthworms that provide the plants with nutrients. I place plants that require a lot of warmth, like orange or kiwi trees, in these 'sun traps.'"

The fundamental principle is to take advantage of the ways in which plants, animals, and the physical environment interact. After Holzer took over the operation from his father in 1962, his first step was to radically change the layout of the sloping fields, creating terraces with integrated ponds. He and his wife, Veronika, have run the farm ever since. They care for the trees according to their own very specific philosophy:

> They are all self-sufficient trees—they don't need to be pruned. Because when you prune trees—and I'm saying this as a trained nursery worker—you have to keep pruning them forever. They become dependent and addicted to it. Trees like that would have no chance up here.

His success speaks for itself: he sells trees to buyers in Germany and Scotland, and year after year visitors flock to the Krameterhof in the fall to harvest fruit.

Just a stone's throw away from Holzer's farm—at least for travelers who think in North American dimensions—is South Tyrol and Italy's northernmost olive orchard. A number of different olive varieties grow here. But the overall impression doesn't match the mental picture that the words "olive orchard" normally evoke. The trees—there are several hundred of them—grow on a mountainside, in the niches of a nearly vertical porphyry rock face. Harvesting the olives is obviously quite a challenge. The location on the bank of the Eisack river was chosen because it is sheltered and especially warm.

Unterganzner, the farm where the olives grow, perches above the town of Bolzano at a point 935 feet (285 meters) above sea level. It has belonged to the same family since 1629. Olives were first grown here in the 1980s, when owner Josephus Mayr fulfilled a dream for himself and planted the trees. Many of his neighbors thought the idea was foolish, and they were even more certain when Mayr lost thousands of olives in the harsh winter of 1986. But now he harvests about two tons of olives each year. Thanks to climate change, they ripen ten to fourteen days earlier than they did a few decades ago. Mayr is proud of the well-rounded taste of his olive oil, which goes down smoothly. At the same time, the olive orchard is really just a hobby; his main operation is a vineyard run according to traditional methods, without the use of pesticides or synthetic fertilizers. But he enjoys the sight of the evergreen olive trees and the contrast they create in the winter against the leafless vines. Apples, figs, nuts, and chestnuts grow on the slopes as well. But at this point that should hardly be surprising.

Fruit trees can grow in other regions where you might not expect them. Surprisingly, the banks of some Norwegian fjords offer an almost ideal climate for cultivation, mostly of apples used to make cider. The country's longest and deepest fjord, the Sognefjord, is a good example. The short vegetation period there is compensated for by long days in summer and a particular microclimate—a "climate oasis," so to speak—that protects the blossoms against frost. Of course, none of these advantages would create suitable conditions for fruit farming without the warming influence of the Gulf Stream. After all, the Sognefjord is on the same latitude as the southern tip of Greenland, the Yukon Territory, and Anchorage.

In the chapter on cherries we saw how bees play an important role in orchards. Some fruit growers in Andalusia go even further to integrate insects into their farms. They have developed a clever method for combating aphids: setting up wooden boxes next to the rows of trees as "hotels" for spiders, the aphid's

natural predators. In this way, they can at least partially reduce the amount of insecticide they use—and fewer insecticides means more beneficial insects of all kinds.

If farmers and orchard operators who bring back traditional fruit-growing methods are "rebels," what should we call the urban activists who are literally taking orcharding to the streets in cities around the world? Dreaming of metropolitan fruit forests, guerrilla grafters splice fruit-bearing scions onto city trees originally bred to be purely ornamental. This practice is technically illegal in some locations because fallen fruit can pose a risk on pedestrian pathways and attract unwanted rodents and insects. It may even be considered vandalism. Guerrilla grafters defend themselves by pointing out that they take care of the trees to make sure they flourish and don't cause any harm. A "tamer" form of bringing fruit directly to neighborhoods can be found in the many community or urban orchards that have sprung up in cities such as San Francisco, Los Angeles, Philadelphia, Vancouver, and others beyond North America.

There are many reasons to celebrate the return of orchards to their rightful place as natural and cultural assets. The widespread longing for orchards and fruits "like they used to be" surely

springs from the fact that so much of our world has been carved up and paved over. Nature has been driven back or destroyed completely and her once-familiar features have faded away. And so orchards where the old trees still stand are considered precious resources. Even trees that have died are sometimes purposefully left in place to create a place for insects, spiders, centipedes, and other tiny creatures to thrive. Wild hedges, piles of branches and stones, and even unused patches of ground provide homes for larger animals like foxes, which pay for their keep by controlling small rodents that might damage the trees' roots or bark.

Here and there traditional celebrations of fruit blossoms are finding new fans who flock to the orchards to remember "the good old days." Beltane, a festival in Great Britain and Ireland with roots in the distant pagan past, has a special role in building awareness of the year's recurring cycle in the orchard: it takes place in early May and begins with bonfires of wood from trees that are considered sacred. Whether for individuals or communities, the gifts of smaller, slower orchards—past and present—go far beyond fruit alone.

A New Beginning

OASES WITH TOWERING DATE palms and smaller fruit trees growing beneath them could have constituted the world's first "orchards." Some exist to this day, radically altered in certain respects and remarkably untouched in others. It's not always possible to determine where the original oases began and ended. And in essence life there has changed very little, although modern developments have certainly left their traces. The water lines are made of different materials than they were four or five thousand years ago, of course, and the work is organized more efficiently. Most of the workers maintaining the date groves of the Arabian Peninsula today are from India, Bangladesh, and Pakistan. They cross the desert with jeeps rather than camels. The pumps are more effective and the water brought up from the depths—and sometimes churned out by desalination plants on the coast—has caused the area cultivated around the oases to spread dramatically.

Dates continue to be a staple for many people. Nefta and Tozeur, oases of Bled el Djerid in the south of Tunisia, lie in a relatively accessible, varied landscape near the Algerian border. They are sheltered from the north winds and contain a large number of natural springs. In addition to date palms, the oases there contain olive, orange, fig, apricot, and peach trees, along with grapevines. Irrigation takes place through a clever system of dams and channels. Traditionally the dams were built of palm trunks with deep, long cuts in them. The number of these cuts determined the amount of water that would flow over the dam in a certain amount of time. In Palmeral de Elche, an oasis of palm trees in southwestern Spain, up to two thousand date palms grow to this day. With a little imagination, you can feel as though you're in an Arabian garden there. The oldest palms are believed to be three hundred years old and many of the trees tower 130 feet (40 meters) into the sky. Each year the palms in Palmeral de Elche produce about two thousand tons of dates.

It is valuable to imagine what our fruit used to be like and to think about how many hands all the seeds, twigs, and trunks must have passed through and what distances they traveled—geographically and through time. Those who plant fruit trees do so not just for themselves but as an investment in the future. In this regard, creating an orchard is a forward-looking project that links generations. And we can also sense this transcending of time in the natural cycle of the trees, which corresponds so closely to the lives of the people who dwell among or near them. Robert Pogue Harrison, a scholar of romance literature at Stanford University, has written about the high moral endeavor that a garden, or in this case, an orchard, represents:

previous page
Peach picking in
Greece, ca. 1960.

*A humanly created garden comes into being in and through
time. It is planned by the gardener in advance, then it is seeded
or cultivated accordingly, and in due time it yields its fruits or
intended gratifications. Meanwhile the gardener is beset by
new cares day in and day out. For like a story, a garden has its
own developing plot, as it were, whose intrigues keep the care-
taker under more or less constant pressure. The true gardener is
always "the constant gardener."*

I grew up far from the country in the middle of Berlin. But
from my earliest childhood I had the great good fortune to spend
the months from spring until fall surrounded by the fruit trees
around our little house. Currant, raspberry, and gooseberry
bushes stood near my beloved swing, out front near the fence to
the neighboring plot. There were bushes with wild strawberries,
too. The fruit was much smaller and more tapered than the strawberries you can buy today. It was less sweet, but with a far more
intense aroma. The little bushes did not suffer the fate of commercially farmed strawberry plants, which are pulled up after the
harvest. We let them stay and each year they developed their fruit
anew. In addition to the berry bushes, we also had several apple,
pear, and plum trees. In the late summer we gathered the ripe
fruit from the branches with a picker—a combination of tongs
and a bag at the end of a long pole. The timing was always perfect: precisely before all that luscious bounty would have fallen on
its own. The trees were tall and we used an unsteady wooden lad-
der. As a young boy, climbing it was always a dizzying adventure
for me.

Once the harvest was brought in, my parents—surrounded
by swarms of wasps—spent several entire weekends cooking the
fruit in a steam juicer, funneling the juice into bottles through a

hose, and carefully sealing the bottles with red rubber plugs. Each year we had quite a juice collection to enjoy during the winter. I remember a large elderberry bush behind the house with fruit that exuded an almost pungently sweet fragrance and stained the terrace with dark blotches. There were even grapes growing in the dense thicket behind our little summerhouse, but they were so small and sour that only the birds enjoyed them.

During the ten years that we had this garden, we added new plants. We were never able to find out who had planted the original fruit trees and bushes. But they were certainly a few decades old and had likely originated after the Second World War. It never occurred to us to hire a "fruit detective" who could precisely identify the plants that so faithfully supplied us year after year. We were happy to live with this minor mystery. In exchange, we had the kind of garden that sometimes overwhelms its owners with so much fruit that they hang bags bursting with apples and pears on the fence as gifts for passersby.

As I was completing this book, I learned about an astounding research project that helps illustrate how early wild fruit could have spread and taken up residence elsewhere. Using GPS transmitters, scientists showed how at least one African species of flying foxes—those large members of the bat family—spread the seeds of fruit trees, helping to foster the regrowth of forests on cleared land. The seeds come from dates, mangoes, and other fruit. Flying foxes behave a lot like migratory birds: they cross forest boundaries and open landscapes and can spread the fruit up to forty-five miles (seventy-five kilometers) away. Imagine the flying foxes of a place like Accra, in Ghana, setting off at sunset to search for fruit beyond the city's borders. Once they have eaten their fill, the seeds remain in their digestive system for one to eight hours and are ejected somewhere on the flight back.

This phenomenon is in no way limited to a small region. These flying foxes can be found from the Ivory Coast on the Atlantic to Kenya on the Indian Ocean. According to biologist Dina Dechmann, of Germany's Max Planck Institute for Ornithology Research and the University of Konstanz, they play a central role in fruit tree survival. "Among the plants spread by the flying foxes are fast-growing trees that act as pioneer species, creating the right environment for other tree species to take root and grow," she says. Working with her Swedish colleague Mariëlle van Toor, from Linnaeus University in Kalmar, Dr. Dechmann has substantiated her observations with concrete numbers. In each of their nightly flights, the estimated 152,000 flying foxes living in Accra distribute 338,000 seeds over a large area. In just a single year they can fill nearly two thousand acres (eight hundred hectares) in Ghana alone with fast-growing trees that bear the types of fruit they eat. Wild-growing fruit forests, if you will, springing to life without the intervention of a single human hand.

Sources for Quotations and Specific Research Cited

Prologue: The Seeds for This Book

"When man migrates…" Henry David Thoreau, *Wild Fruits*, ed. Bradley Dean (New York: Norton, 2001).

The essay that was the impetus for this book is George Willcox, "Les fruits au Proche-Orient avant la domestication des fruitiers," in Marie-Pierre Ruas, ed., *Des fruits d'ici et d'ailleurs: Regards sur l'histoire de quelques fruits consommés en Europe* (Paris: Omni-science, 2016).

"as far as plants are concerned…" (and following quote) Ahmad Hegazy and Jon Lovett-Doust, *Plant Ecology in the Middle East* (Oxford: Oxford University Press, 2016).

"It has consisted in always cultivating…" Charles Darwin, *On the Origin of Species* (London: John Murray, 1859).

1 Before There Were Orchards

Articles mentioned in this chapter:

Alexandra DeCasien, Scott A. Williams, and James P. Higham, "Primate Brain Size Is Predicted by Diet but Not Sociality," *Nature Ecology & Evolution* 1, no. 5 (March 2017).

Nathaniel J. Dominy et al., "How Chimpanzees Integrate Sensory Information to Select Figs," *Interface Focus* 6, no. 3 (June 2016).

Mordechai E. Kislev, Anat Hartmann, and Ofer Bar-Yosef, "Early Domesticated Fig in the Jordan Valley," *Science* 312, no. 5778 (July 2006).

"Your wife will be like a fruitful vine…" *The Bible, New International Edition*, Psalm 128:3.

"Olives were domesticated before…" Mort Rosenblum, *Olives: The Life and Lore of a Noble Fruit* (Bath: Absolute Press, 1977).

2 The Rustle of Palm Leaves

Some of the information in this chapter is based on the informative though difficult-to-find book by Warda H. Bircher, *The Date Palm: A Friend and Companion of Man* (Cairo: Modern Publishing House, 1995).

"the loftiest and most stately of all vegetable forms…" Alexander von Humboldt, *Views of Nature: Or, Contemplations on the Sublime Phenomena of Creation* (London: Henry G. Bohn, 1850).

"They are surrounded by dry walls…" Quoted in Berthold Volz ed., *Geographische Charakterbilder aus Asien* (Leipzig: Fuess, 1887) (in translation).

3 Gardens of the Gods

"Like her teeth my seeds…" Quoted in Maureen Carroll, *Earthly Paradises: Ancient Gardens in History and Archaeology* (Los Angeles: Getty Publications, 2003).

"I dug out a canal from the Upper Zab…" (and following quote) Quoted in Stephanie Dalley, "Ancient Mesopotamian Gardens and the Identification of the Hanging Gardens of Babylon Resolved," *Garden History* 21, no. 1 (Summer 1993).

"They entered through a vaulted gateway…"
Quoted in Muhsin Mahdi, ed., *The Arabian Nights* (New York: Norton, 2008).

"But they are gardens of trees…" Vita Sackville-West, *Passenger to Teheran*, (London: Hogarth Press, 1926).

"The Lord God took the man…" *The Bible, New International Edition*, Genesis 2:15.

4 Not Far From the Tree

"We saw, for the first time…" John Selborne, "Sweet Pilgrimage: Two British Apple Growers in the Tian Shan," *Steppe: A Central Asian Panorama* 9 (Winter 2011).

"We found it to be enclosed by a high wall…" Quoted in Jonas Benzion Lehrman, *Earthly Paradise: Garden and Courtyard in Islam* (Berkeley: University of California Press, 1980).

A comprehensive source on the origins of the apple is Barrie E. Juniper and David J. Mabberley, *The Story of the Apple* (Portland: Timber Press, 2006). Please also see the recent book by Robert N. Spengler III, *Fruit From the Sands: The Silk Road Origins of the Fruits We Eat* (Berkeley: University of California Press, 2017).

5 Studying the Classics

"Close to the gates a spacious garden lies…" Homer, *The Odyssey*, trans. Samuel Butler (originally published in 800 BC; Butler's translation published in London: A. C. Fifield, 1900) (accessed online through Project Gutenberg).

"The three essential 'resonances' were…" Margaret Helen Hilditch, *Kepos: Garden Spaces in Ancient Greece: Imagination and Reality* (doctoral dissertation, University of Leicester, 2015),

https://pdfs.semanticscholar.org/540b/ 8fb60465ccd7e92a3c3feba9234f6eb4ee1.pdf.

"nothing more excellent or valuable was ever granted…" Quoted in Archibald F. Barron, *Vines and Vine Culture* (London: Journal of Horticulture, 1883).

"Many and all manner of nations dwell in the Caucasus…" (and following quote) Herodotus, *The Persian Wars*, trans. A. D. Godley (originally published ca. 430 BCE; Godley's translation published in Cambridge, Mass.: Harvard University Press, 1920).

"so many fields, so many types…" Theophrastus, *Enquiry Into Plants* and *On the Causes of Plants*, trans. Arthur F. Hort (originally published ca. 350–287 BCE. Hort's translation published in Cambridge, Mass.: Harvard University Press, 1916).

"No less, meanwhile…" Vergil, *Georgica*, trans. Theodore Chickering Williams (originally published ca. 37–29 BCE; Williams's translation published in Cambridge, Mass.: Harvard University Press, 1915).

"Plant the pear tree in the autumn before winter comes…" (and following quote) Lucius Junius Moderatus Columella, *On Agriculture*, trans. Harrison Boyd Ash (originally published in 1559; Ash's translation published in Cambridge, Mass.: Harvard University Press, 1941–55).

"In the matter of fruit-trees…" Pliny the Elder, *Natural History*, book 17, trans. Harris Rackham (originally published ca. 77 CE; Rackham's translation published in Cambridge, Mass.: Harvard University Press, 1938–63).

"Next to the *gestatio*…" (and following quote) Pliny the Younger, *Letters*, trans. William Melmoth (Cambridge, Mass.: Harvard University Press, 1963).

"It is thought that..." Marcus Terentius Varro, *On Agriculture*, originally published in 37 BCE; trans. William Davis Hooper, rev. Harrison Boyd Ash, (Hooper's translation published in Cambridge, Mass.: Harvard University Press, 1934).

"The climate is wretched..." Tacitus, *The Agricola and the Germania*, trans. Harold Mattingly (Harmondsworth, England: Penguin Books, 1948).

6 Earthly Paradises

Information on the monastery of St. Gall can be found in Walter Horn and Ernest Born, *The Plan of St. Gall: A Study of the Architecture and Economy of, and Life in a Paradigmatic Carolingian Monastery* (Berkeley and Los Angeles: University of California Press, 1979).

"You sit in the fenced enclosure of your small garden..." Walafrid Strabo, *On the Cultivation of Gardens: A Ninth Century Gardening Book* (San Francisco: Ithuriel's Spear, 2009).

"There was a garden for the monks..." Gregory of Tours, "The Lives of the Fathers," ca. 14 CE, in Gregory of Tours, *Lives and Miracles*, ed. and trans. Giselle de Nie, Dumbarton Oaks Medieval Library 39 (Cambridge, Mass.: Harvard University Press, 2015).

"Many fruit-bearing trees of various species..." Quoted in Stephanie Hauschild, *Das Paradies auf Erden. Die Gärten der Zisterzienser* (Ostfildern: Jan Thorbecke Verlag, 2007) (in translation).

Sonia Lesot's books about the garden she bought with Patrice Taravella: *Les jardins du prieuré Notre-Dame d'Orsan* (Arles: Actes Sud, 1999), and (with Henri Gaud) *Orsan: Des jardins d'inspiration monastique médiévale* (Arles: Editions Gaud, 2003).

"What is more beautiful than the quincunx..." (and following quote) Quintilian, *The Institutio Oratoria of Quintilian*, vol. 3, trans. Harold Edgeworth Butler (Cambridge, Mass.: Harvard University Press, 1959–63).

"Fragrant and perfectly ripened pears..." (and following quote) Ibn Butlan, *The Four Seasons of the House of Cerutti* (New York: Facts on File, 1984).

"It was sprinkled all over..." Giovanni Boccaccio, *The Decameron* (New York: Norton, 2015).

"goodliest trees laden with fairest fruit..." Quoted in Paul A. Underwood, *The Fountain of Life in Manuscripts of Gospels* (Washington, DC: Dumbarton Oaks Papers, 1950).

"The world is a great library..." Ralph Austen, *The Spiritual Use of an Orchard or Garden of Fruit Trees* (Oxford: Printed for Thos. Robinson, 1653).

"A well-contriv'd Fruit-Garden is an Epitome of Paradise itself..." Stephen Switzer, *The Practical Fruit-Gardener* (London: Thomas Woodward, 1724).

"Of all other delights on earth..." (and two following quotes) William Lawson, *A New Orchard and Garden* (London: n.p., 1618).

7 Pears for the Sun King

"It must be confessed that..." Jean-Baptiste de La Quintinie, *The Complete Gard'ner: Or, Directions for Cultivating and Right Ordering of Fruit-Gardens and Kitchen-Gardens* (London: Andrew Bell, 1710).

"Does the musk or amber exist..." René Dahuron, *Nouveau traité de la taille des arbres fruitiers* (Paris: Charles de Sercy, 1696) (in translation).

"The apple tree is the most necessary and valuable of all…" Charles Estienne and Jean Liebault, *L'agriculture et maison rustique* (Paris: Jacques du Puis, 1564) (in translation).

"eight hours before Christmas…" Quoted in Florent Quellier, *Des fruits et des hommes: L'arboriculture fruitière en Île-de-France (vers 1600–vers 1800)* (Rennes: Presses Universitaires des Rennes, 2003) (in translation).

"A fruit at full ripeness can only be healthy…" de La Bretonnerie, *L'école du jardin fruitier* (Paris: Eugène Onfroy, 1784) (in translation).

8 Moving North

"Fruit gathered too timely will taste of the wood…" Thomas Tusser, *Five Hundred Pointes of Good Husbandrie* (London: Lackington, Allen, and Co., 1812).

"to convey the produce…" Quoted in Susan Campbell, "The Genesis of Queen Victoria's Great New Kitchen Garden," *Garden History* 12, no. 2 (Autumn 1984).

"Kitchen-gardens are not as well-kept as ours…" François de La Rochefoucauld, *A Frenchman's Year in Suffolk,* trans. Norman Scarfe (Woodbridge: Boydell Press, 2011).

"Great abundance of as good vines, peaches, apricots…" Quoted in Sandra Raphael, *An Oak Spring Pomona: A Selection of the Rare Books on Fruit in the Oak Spring Garden Library* (Upperville, Virginia: Oak Spring Garden Library, 1990).

The information on the third Duke of Hamilton comes from Rosalind K. Marshall, *The Days of Duchess Anne* (Edinburgh: Tuckwell Press, 2000).

On orchards in Scotland see also Forbes W. Robertson, "A History of Apples in Scottish Orchards," *Garden History* 35, no. 1 (Summer 2007).

9 Orchards for the Masses

The information on the orchards and forests around Paris is from Florent Quellier, *Des fruits et des hommes: L'arboriculture fruitière en Île-de-France (vers 1600–vers 1800)* (Rennes: Presses Universitaires des Rennes, 2003).

"of freely growing trees is superior to all others…" (and following quote) Henri-Louis Duhamel du Monceau, *Traité des arbres fruitiers* (Paris: Jean Desaint, 1768) (in translation).

"the cultivation of fruit trees…" (and following quotes) Quoted in Rupprecht Lucke, Robert Silbereisen, and Erwin Herzberger, *Obstbäume in der Landschaft* (Stuttgart: Ulmer, 1992) (in translation).

"trees that are highly suitable for lining roads…" (and other quotes) Johann Caspar Schiller, *Die Baumzucht im Großen* (Neustrelitz: Hofbuchhandlung, 1795) (in translation).

"the harvest was turned into cider, dried fruit, or brandy…" (and other anonymous quotes in this chapter) Quoted in Rupprecht Lucke et al., *Obstbäume in der Landschaft* (in translation).

"It was highly dangerous…" Quoted in Eric Robinson, "John Clare: 'Searching for Blackberries,'" *The Wordsworth Circle* 38, no. 4 (Autumn 2007).

"The bawling song of solitary boys…" (and following quote) John Clare, *The Shepherd's Calendar* (Oxford: Oxford University Press, 2014).

10 Cherry Picking

"On the 25th I will travel to Amalthea..." Max Hein, ed., *Briefe Friedrichs des Grossen* (Berlin: Reimar Hobbing, 1914) (in translation).

"A host of bards have raised their song..." Anna Louisa Karsch, "Lob der schwarzen Kirschen," in *Auserlesene Gedichte* (Berlin: George Ludewig Winter, 1764) (in translation).

"There is another circumstance in which I am very particular..." Joseph Addison, n.t., *The Spectator*, no. 477 (September 6, 1712).

11 Pucker Up

"Do you know the land where the lemon-trees grow..." Johann Wolfgang von Goethe, *Wilhelm Meister's Apprenticeship*, trans. Eric A. Blackall (Princeton: Princeton University Press, 1995).

"In this profound and delicious solitude..." Jean-Jacques Rousseau, *The Confessions*, trans. J. M. Cohen (London: Penguin, 1953).

"In the entire garden there are neither plants nor trees..." (and following quotes) Jean-Baptiste de La Quintinie, *The Complete Gard'ner: Or, Directions for Cultivating and Right Ordering of Fruit-Gardens and Kitchen-Gardens* (London: Andrew Bell, 1710).

"An autumn day, pavilion in the wilds..." Du Fu, *The Poetry of Du Fu* (Berlin: De Gruyter, 2016).

"And though that place is 1,000 kos distant..." Nuru-d-din Jahangir Padshah, *The Tuzuk-i-Jahangiri: or, Memoirs of Jahangir*, trans. Alexander Rogers, ed. H. Beveridge (first published ca. 1609; Rogers's translation published in Ghazipur: 1863; Beveridge's revised edition in London: Royal Asiatic Society, 1909).

"I have left no stone unturned..." (and following quote) Emanuel Bonavia, *The Cultivated Oranges, Lemons etc. of India and Ceylon* (London: W. H. Allen & Co., 1888).

"As I intended to make some observations..." Quoted in Carsten Schirarend and Marina Heilmeyer, *Die Goldenen Äpfel. Wissenswertes rund um die Zitrusfrüchte* (Berlin: G + H Verlag, 1996).

"Bear me, Pomona, to thy citron grove..." James Thomson, *The Seasons and the Castle of Indolence* (London: Pickering, 1830).

"We passed Limone..." (and following quote) Johann Wolfgang von Goethe, *Italian Journey*, trans. W. H. Auden and Elizabeth Mayer (London: Penguin, 1970).

"We like to be out in nature so much..." Friedrich Nietzsche, *Human, All Too Human* (London: Penguin, 1994).

A comprehensive treatment of Nietzsche's stay in southern Italy can be found in Paolo D'Iorio and Sylvia Mae Gorelick, *Nietzsche's Journey to Sorrento: Genesis of the Philosophy of the Free Spirit* (Chicago: The University of Chicago Press, 2016).

"Have you ever slept, my friend..." Guy de Maupassant, "The Mountain Pool," *Original Short Stories*, vol. 13, trans. A. E. Henderson, Louise Charlotte Garstin Quesada, Albert Cohn McMaster, ed. David Widger, accessed through The Gutenberg Project, http://www.gutenberg.org/files/28076/28076-h/28076-h.htm.

12 As American as Apple Pie

"The Country is full of gallant Orchards..." John Hammond, *Leah and Rachel; or, The Two Fruitfull Sisters, Virginia and Mary-Land* (London: Mabb, 1656).

"having planted in the fall a new apple orchard…" (and following quotes) Hector St. John de Crèvecoeur, *Sketches of Eighteenth Century America: More Letters From an American Farmer* (New Haven: Yale University Press, 1925).

"Grafted 5 others of the same Cherry's…" *The Diaries of George Washington*, vol. 1, 11 March 1748–13 November 1765, ed. Donald Jackson (Charlottesville: University Press of Virginia, 1976).

"the best cyder apple existing…" (and following quotes) Quoted in Peter J. Hatch, *The Fruits and Fruit Trees of Monticello* (Charlottesville: University of Virginia Press, 1998).

"There is comfort in a good orchard…" Quoted in Eric Rutkow, *American Canopy: Trees, Forests and the Making of a Nation* (New York: Scribner, 2012).

"the apple is our national fruit…" Ralph Waldo Emerson, *The Journals and Miscellaneous Notebooks of Ralph Waldo Emerson: 1848–1851* (Boston: Harvard University Press, 1975).

"the pie is an English tradition…" Harriet Beecher Stowe, *Oldtown Folks* (Boston: Fields, Osgood & Co., 1869).

"They can improve the apple in that way…" Quoted in Howard Means, *Johnny Appleseed: The Man, the Myth, the American Story* (New York: Simon and Schuster, 2012).

"A railroad line used to run up…" Philip Roth, *American Pastoral* (Boston: Houghton Mifflin, 1997).

Much of the information on American developments related to citrus culture here is based on the excellent book by Pierre Laszlo, *Citrus: A History* (Chicago: University of Chicago Press, 2007).

13 Orchards Unbound

"In many cases, human activity…" Charles M. Peters, *Managing the Wild: Stories of People and Plants and Tropical Forests* (New Haven: Yale University Press 2018).

"In this flower garden Montezuma did not allow…" (and following quote) Quoted in Patrizia Granziera, "Concept of the Garden in Pre-Hispanic Mexico," *Garden History* 29, no. 2 (Winter 2001).

"Straight up from the water the forest rises like a wall…" Quoted in Nigel Smith, *Palms and People in the Amazon* (Cham: Springer, 2015). The following quote ("Many areas of the Amazon…") is also from this book.

"The Count ordered them to be fetched…" (and other quotes by Caspar Barlaeus) Quoted in Maria Angélica da Silva and Melissa Mota Alcides, "Collecting and Framing the Wilderness: The Garden of Johan Maurits (1604–79) in North-East Brazil," *Garden History* 30, no. 2 (Winter 2002).

"The beautiful effect of the landscape…" Ernst Haeckel, *A Visit to Ceylon* (New York: Peter Eckler, 1883).

"More than 2000 years of agricultural activities…" Karin Hochegger, *Farming Like the Forest: Traditional Home Garden Systems in Sri Lanka* (Weikersheim: Margraf Verlag, 1998).

"Gardening among the Singalese…" John Davy, *An Account of the Interior of Ceylon and of Its Inhabitants* (London: Longman, Hurst, Rees, Orm, and Brown, 1821).

"[The tree was] tall with tiny yellow berries…" Michael Ondaatje, *Running in the Family* (New York: Vintage, 1993).

"A cord is fastened round the monkey's waist…" Robert W. C. Shelford, *A Naturalist in Borneo* (London: T. F. Unwin, 1916).

14 Pomological Gentlemen

"promoting fruit culture in British
dominions..." *The Horticulturist and
Journal of Rural Art and Rural Taste*, vol. 4
(Albany: L. Tucker, 1854).

"to graffe many sortes of Apples on one Tree..."
(and following quotes) Leonard Mascall,
*A Booke of the Arte and Maner, Howe to Plant
and Graffe all Sortes of Trees...* (London: By
Henrie Denham, for John Wight, 1572).

"alternate in close proximity or worse be completely mixed together..." (and following
quotes) Carl Samuel Häusler, *Aphorismen*
(Hirschberg: C. W. J. Krahn, 1853).

"across the Atlantic Ocean..." (and following
quote) Thomas Skip Dyot Bucknall, *The
Orchardist, or, A System of Close Pruning and
Medication, for Establishing the Science of
Orcharding* (London: G. Nicol, 1797).

"Many useful manures may be obtained..."
William Salisbury, *Hints Addressed to
Proprietors of Orchards, and to Growers of
Fruit in General* (London: Longman, Hurst,
Rees, Orme, and Brown, 1816).

According to my research, there is no book
on Korbinian Aigner in English. The only
currently available resource on his work in
German is Peter Brenner, *Korbinian Aigner:
Ein bayerischer Pfarrer zwischen Kirche, Obst-
garten und Konzentrationslager* (Munich:
Bauer-Verlag, 2018).

"I have no faith in the selected lists of pomo-
logical gentlemen..." (and other quotes)
Henry David Thoreau, "Wild Apples," *The
Atlantic*, November 1862.

15 Orchards of the Senses

"Come let us watch the sun go down..." Rainer
Maria Rilke, "Der Apfelgarten," in *Selected
Poems*, trans. Albert Ernest Flemming (New
York: Routledge, 2011).

"apricots are plums concealed beneath a
peach's coat..." Quoted in Edward Bunyard,
The Anatomy of Dessert (London: Dulau &
Co., 1929).

"Some botanists consider the myrtle..." Vita
Sackville-West, *In Your Garden* (London:
Frances Lincoln, 2004).

"The olive tree, what a brute..." Quoted in
Derek Fell, *Renoir's Garden* (London: Frances Lincoln, 1991).

"With an apple, I will..." (*Avec une pomme, je
veux étonner Paris!*) Quoted in Gustave
Geffroy, *Claude Monet, sa vie, son temps, son
œuvre* (Paris: G. Crès, 1924).

"Some keep the Sabbath going to Church..."
Emily Dickinson, *Poems*, Mabel Loomis
Todd and Thomas Wentworth Higginson,
eds. (Boston: Roberts Brothers, 1890).

"One chief grace that adorns an Orchard..."
William Lawson, *A New Orchard and Garden*
(London: n.p., 1618).

"the roar of multitudinous wheels..." John
James Platt, "The Pleasures of Country Life,"
in *A Return to Paradise and Other Fly-Leaf
Essays in Town and Country* (London: James
Clarke & Co., 1891).

"Miranda slept in the orchard..." Virginia
Woolf, "In the Orchard," *The Criterion* (London: R. Cobden-Sanderson, 1923).

"the advance of the year was so rapid..."
(and following quotes) Quoted in Anne
Scott-James, *The Language of the Garden:
A Personal Anthology* (New York: Viking,
1984).

"From the middle of the summer…" Carl
Larsson, *Our Farm* (London: Methuen
Children's Books, 1977).

"I see no reason for placing the kitchen
garden…" William Cobbett, *The English
Gardener* (London: A. Cobbett, 1845).

16 Returning to the Wilder Ways of Fruit

"the most promising and important apple…"
David Karp, "Beyond the Honeycrisp Apple,"
New York Times, November 3, 2015.

"Those who have tasted them speak of them
reverently…" Mentioned in Clarissa
Hyman, "Forbidden Fruit," *Times Literary
Supplement*, December 23 and December
30, 2016.

"The last of my Sun Crest peaches will be dug
up…" David Mas Masumoto, *Epitaph for
a Peach: Four Seasons on My Family Farm*
(New York: HarperCollins, 1996).

"The stones 'sweat' and water condenses
underneath them," (and following quotes)
Quoted in Florianne Koechlin, *Pflanzen-
Palaver. Belauschte Geheimnisse der botani-
schen Welt* (Basel: Lenos Verlag, 2008) (in
translation).

Epilogue: A New Beginning

"A humanly created garden…" Robert Pogue
Harrison, *Gardens: An Essay on the Human
Condition* (Chicago: University of Chicago
Press, 2008).

"Among the plants spread by the flying foxes…"
Quoted in "Fruit Bats Are Reforesting
African Woodlands," press release of the
Max Planck Society, April 1, 2019.

Further Reading

Attlee, Helena. *The Land Where Lemons Grow: The Story of Italy and Its Citrus Fruit*. London: Penguin, 2014.

Barker, Graeme and Candice Goucher (eds.). *The Cambridge World History*. Vol. 2, *A World With Agriculture, 12,000 BCE–500 CE*. Cambridge: Cambridge University Press, 2015.

Beach, Spencer Ambrose. *The Apples of New York*. Albany: J. B. Lyon, 1903.

Bennett, Sue. *Five Centuries of Women and Gardens*. London: National Portrait Gallery, 2000.

Biffi, Annamaria and Susanne Vogel. *Von der gesunden Lebensweise. Nach dem alten Hausbuch der Familie Cerruti*. München: BLV Buchverlag, 1988.

Bircher, Warda H. *The Date Palm: A Friend and Companion of Man*. Cairo: Modern Publishing House, 1995.

Blackburne-Maze, Peter and Brian Self. *Fruit: An Illustrated History*. Richmond Hill: Firefly Books, 2003.

Boccaccio, Giovanni. *The Decameron*. New York: Norton, 2014.

Brosse, Jacques. *Mythologie des arbres*. Paris: Plon, 1989.

Brown, Pete. *The Apple Orchard: The Story of Our Most English Fruit*. London: Particular Books, 2016.

Candolle, Alphonse Pyrame de. *Origin of Cultivated Plants*. New York: D. Appleton & Co., 1883.

Carroll-Spillecke, M., ed. *Der Garten von der Antike bis zum Mittelalter*. Mainz: Verlag Philipp von Zabern, 1992.

Crèvecoeur, Hector St. John de. *Letters From an American Farmer* and *Sketches of Eighteenth Century America: More Letters From an American Farmer*. New Haven: Yale University Press, 1925.

Daley, Jason. "How the Silk Road Created the Modern Apple." *Smithsonian.com*, August 21, 2017.

Dalley, Stephanie. "Ancient Mesopotamian Gardens and the Identification of the Hanging Gardens of Babylon Resolved." *Garden History* 21, no. 1 (Summer 1993).

Diamond, Jared. *Guns, Germs and Steel: The Fates of Human Societies*. New York: W. W. Norton, 1999.

Fell, Derek. *Renoir's Garden*. London: Frances Lincoln, 1991.

Gignoux, Emmanuel, Antoine Jacobsohn, Dominique Michel, Jean-Jacques Peru, and Claude Scribe. *L'ABCdaire des Fruits*. Paris: Flammarion, 1997.

Gollner, Adam Leith. *The Fruit Hunters: A Story of Nature, Adventure, Commerce and Obsession*. London: Souvenir Press, 2009.

Harris, Stephen. *What Have Plants Ever Done for Us? Western Civilization in Fifty Plants*. Oxford: Bodleian Library, 2015.

Harrison, Robert Pogue. *Gardens: An Essay on the Human Condition*. Chicago: University of Chicago Press, 2008.

Hauschild, Stephanie. *Akanthus und Zitronen. Die Welt der römischen Gärten*. Darmstadt: Philipp von Zabern, 2017.

——. *Das Paradies auf Erden. Die Gärten der Zisterzienser*. Ostfildern: Jan Thorbecke Verlag, 2007.

——. *Der Zauber von Klostergärten*. München: Dort-Hagenhausen-Verlag, 2014.

Hegazy, Ahmad and Jon Lovett-Doust. *Plant Ecology in the Middle East*. Oxford: Oxford University Press, 2016.

Hehn, Victor. *Cultivated Plants and Domesticated Animals in Their Migration From Asia to Europe*. London: Swan Sonnenschein & Co., 1885.

Heilmeyer, Marina. *Äpfel fürs Volk: Potsdamer Pomologische Geschichten*. Potsdam: vacat verlag, 2007.

——. *Kirschen für den König: Potsdamer Pomologische Geschichten*. Potsdam: vacat verlag, 2008.

Hirschfelder, Hans Ulrich, ed. *Frische Feigen: Ein literarischer Früchtekorb*. Frankfurt am Main: Insel Verlag, 2000.

Hobhouse, Penelope. *Plants in Garden History: An Illustrated History of Plants and Their Influences on Garden Styles—From Ancient Egypt to the Present Day*. London: Pavilion, 1992.

Hochegger, Karin. *Farming Like the Forest: Traditional Home Garden Systems in Sri Lanka*. Weikersheim: Margraf Verlag, 1998.

Horn, Walter and Ernest Born. *The Plan of St. Gall: A Study of the Architecture and Economy of, and Life in a Paradigmatic Carolingian Monastery*. Berkeley and Los Angeles: University of California Press, 1979.

Janson, H. Frederic. *Pomona's Harvest: An Illustrated Chronicle of Antiquarian Fruit Literature*. Portland: Timber Press, 1996.

Jashemski, Wilhelmina F. *The Gardens of Pompeii: Herculaneum and the Villas Destroyed by Vesuvius*. New Rochelle, New York: Caratzas Brothers Publishers, 1979.

Juniper, Barrie E. and David J. Mabberley. *The Story of the Apple*. Portland: Timber Press, 2006.

Klein, Joanna. "Long Before Making Enigmatic Earthworks, People Reshaped Brazil's Rain Forest." *The New York Times*, February 10, 2017.

Küster, Hansjörg. *Geschichte der Landschaft in Mitteleuropa: Von der Eiszeit bis zur Gegenwart*. Munich: C. H. Beck, 1995.

Larsson, Carl. *Our Farm*. London: Methuen Children's Books, 1977.

Laszlo, Pierre. *Citrus: A History*. Chicago: University of Chicago Press, 2007.

Lawton, Rebecca. "Midnight at the Oasis." *Aeon*, November 6, 2015.

Lucke, Rupprecht, Robert Silbereisen, and Erwin Herzberger. *Obstbäume in der Landschaft*. Stuttgart: Ulmer, 1992.

Lutz, Albert, ed. *Gärten der Welt: Orte der Sehnsucht und Inspiration*. Museum Rietberg Zürich. Cologne: Wienand Verlag, 2016.

Mabey, Richard. *The Cabaret of Plants: Botany and the Imagination*. London: Profile Books, 2015.

Martini, Silvio. *Geschichte der Pomologie in Europa*. Bern: self-pub., 1988.

Masumoto, David Mas. *Epitaph for a Peach: Four Seasons on My Family Farm*. New York: HarperCollins, 1996.

Mayer-Tasch, Peter Cornelius and Bernd Mayerhofer. *Hinter Mauern ein Paradies: Der mittelalterliche Garten*. Frankfurt am Main und Leipzig: Insel Verlag, 1998.

McMorland Hunter, Jane and Chris Kelly. *For the Love of an Orchard: Everybody's Guide to Growing and Cooking Orchard Fruit*. London: Pavilion, 2010.

McPhee, John. *Oranges*. New York: Farrar, Straus & Giroux, 1967.

Müller, Wolfgang. *Die Indianer Amazoniens: Völker und Kulturen im Regenwald*. Munich: C. H. Beck, 1995.

Nasrallah, Nawal. *Dates: A Global History.* London: Reaktion Books, 2011.

Palter, Robert. *The Duchess of Malfi's Apricots and Other Literary Fruits.* Columbia: The University of South Carolina Press, 2002.

Pollan, Michael. *The Botany of Desire: A Plant's-Eye View of the World.* New York: Random House, 2001.

Potter, Jennifer. *Strange Bloom: The Curious Lives and Adventures of the John Tradescants.* London: Atlantic Books, 2008.

Quellier, Florent. *Des fruits et des hommes: L'arboriculture fruitière en Île-de-France (vers 1600–vers 1800).* Rennes: Presses Universitaires des Rennes, 2003.

Raphael, Sandra. *An Oak Spring Pomona: A Selection of the Rare Books on Fruit in the Oak Spring Garden Library.* Upperville, Virginia: Oak Spring Garden Library, 1990.

Roach, Frederick A. *Cultivated Fruits of Britain: Their Origin and History.* London: Blackwell, 1985.

Rosenblum, Mort. *Olives: The Life and Lore of a Noble Fruit.* Bath: Absolute Press, 2000.

Rutkow, Eric. *American Canopy: Trees, Forests and the Making of a Nation.* New York: Scribner, 2012.

Sackville-West, Vita. *In Your Garden.* London: Francis Lincoln, 2004.

——. *Passenger to Teheran.* London: Hogarth Press, 1926.

Schermaul, Erika. *Paradiesapfel und Pastorenbirne. Bilder und Geschichten von alten Obstsorten.* Ostfildern: Jan Thorbecke Verlag, 2004.

Scott, James C. *Against the Grain: A Deep History of the Earliest States.* New Haven: Yale University Press, 2017.

Scott-James, Anne. *The Language of the Garden: A Personal Anthology.* New York: Viking, 1984.

Selin, Helaine, ed. *Encyclopedia of the History of Science, Technology, and Medicine in Non-Western Cultures.* Dordrecht: Springer Science+Business Media, 1997 (entry by Georges Métailié: "Ethnobotany in China," p. 314).

Sitwell, Osbert. *Sing High! Sing Low! A Book of Essays.* London: Gerald Duckworth & Co., 1943.

Smith, J. Russell. *Tree Crops: A Permanent Agriculture.* New York: Harcourt, Brace and Company, 1929.

Smith, Nigel. *Palms and People in the Amazon.* Cham: Springer, 2014.

Sutton, David C. *Figs: A Global History.* London: Reaktion Books, 2014.

Sze, Mai-mai, ed. *The Mustard Seed Garden Manual of Painting.* Princeton: Princeton University Press, 1978.

Thoreau, Henry David. "Wild Apples." *The Atlantic,* November 1862.

——. *Wild Fruits.* ed. Bradley Dean. New York: Norton, 2001.

Young, Damon. *Philosophy in the Garden.* Victoria: Melbourne University Press, 2012.

Zohary, Daniel and Maria Hopf. *Domestication of Plants in the Old World: The Origin and Spread of Cultivated Plants in West Asia, Europe and the Nile Valley.* Oxford: Oxford University Press, 2000.

Illustration Credits

170 Spreading out apples to dry, Nicholson Hollow, Shenandoah National Park, Virginia. Photograph by Arthur Rothstein, 1935. Library of Congress, LC-USF34- 000361-D.

172/173 *American Homestead Autumn*. Lithograph by Currier and Ives, 1868/1869.

175 Carrying crates of peaches from the orchard to the shipping shed, Delta County, Colorado. Photograph by Lee Russell, 1940. Library of Congress, LC-USF33- 012890-M2.

178 Promotional poster for Gilbert Orchards by Cragg D. Gilbert, early 1950s.

179 Interior of Mrs. Botner's storage cellar, Nyssa Heights, Malheur County, Oregon. Photograph by Dorothea Lange, 1939. Library of Congress, LC-USF34- 021432-E.

181 Orange harvest in southern California. Early twentieth-century postcards.

183 A Wagon Load of Grape Fruit, Florida. Early twentieth-century postcard.

186 Jungle in South Asia from Hermann Wagner, *Naturgemälde der ganzen Welt*. Esslingen: Verlag von J. F. Schreiber, 1869.

188 Papaya from George Meister, *Der Orientalisch-Indianische Kunst- und Lust-Gärtner*. Dresden: Meister, 1692.

190 Boy climbing a palm tree by J. Huyot from Jacques-Henri Bernardin de Saint-Pierre, *Paul et Virginie*. Paris: Librairie Charles Tallandier, ca. 1890.

191 Red howler monkey from *Das Buch der Welt*. Stuttgart: Hoffmann'sche Verlags-Buchhandlung, 1858.

193 Tropical fruit from *Meyers Konversations-Lexikon*. Leipzig and Vienna: Bibliographisches Institut, 1896.

195 Pisang (banana) from George Meister, *Der Orientalisch-Indianische Kunst- und Lust-Gärtner*. Dresden: Meister, 1692.

196 Gewatta near Kandy, Sri Lanka. Contemporary photograph by Sarath Ekanayake.

199 Activities on fruit tree plantations in the area around what is now Rio de Janeiro, Brazil, by André Thevet, *Les singularitez de la France antarctique*. Paris: Chez les heritiers de Maurice de la Porte, 1558.

201 Chromalith by W. Koehler after Ernst Haeckel's 1882 painting depicting virgin forest at the Blue River, Kelany-Ganga, Ceylon (now Sri Lanka) from Ernst Haeckel, *Wanderbilder: Die Naturwunder der Tropenwelt. Ceylon und Insulinde*. Gera: Koehler, 1906.

202/203 Pineapple and avocado by J. Huyot from Jacques-Henri Bernardin de Saint-Pierre, *Paul et Virginie*. Paris: Librairie Charles Tallandier, ca. 1890.

204 Wild crab apple, Bowling Green, Pike County, Missouri, 1914, from the U.S. Department of Agriculture Pomological Watercolor Collection. Rare and Special Collections, National Agricultural Library, Beltsville, MD.

206 Garden at harvest time from Eduard Walther, *Bilder zum ersten Anschauungsunterricht für die Jugend*. Esslingen bei Stuttgart: Verlag von J. F. Schreiber, ca. 1890.

209 Figs from George Brookshaw, *Pomona Britannica*. London: Longman, Hurst, 1812.
211 Korbinian Aigner. With kind permission of TUM.Archiv, Technical University Munich.
212 Korbinian Aigner in the company of seminarians, ca. 1910. With kind permission of TUM.Archiv, Technical University Munich.
214 KZ3 or Korbinian's apple by Korbinian Aigner. With kind permission of TUM. Archiv, Technical University Munich.
216 Detail from *Apple Blossoms* (also known as *Spring*) by John Everett Millais, ca. 1856–1859. Lady Lever Art Gallery, Liverpool. Wikimedia Commons.
219 Detail from *In the Orchard* by Sir James Guthrie, 1885–1886. Scottish National Gallery, purchased jointly by the National Galleries of Scotland and Glasgow Life with assistance from the National Heritage Memorial Fund and the Art Fund, 2012. Wikimedia Commons.
221 Olive groves by Vincent van Gogh, 1889/1890. Museum of Modern Art, New York (above: F712, below: F655). Wikimedia Commons.
222 *The Lovers* by Pierre-Auguste Renoir, 1875. National Gallery, Prague, Czech Republic, no. O 3201. Wikimedia Commons.
225 *Lady With Parasol* by Louis Lumière, 1907.
227 Detail from *The Apple Gatherers*, Frederick Morgan, 1880. Wikimedia Commons.
229 Detail from *In the Orchard* by Edmund Charles Tarbell, 1891. Terra Foundation for American Art, Daniel J. Terra Collection. Wikimedia Commons.

Index of People and Places

Numbers in bold indicate illustrations

Euphrates (valley and river), 10, 17, 31;
Nimrud, 31; Tigris, 23, 31, 17, 32, 35; Ur, 17
Mexico: iv, 188, 189; Huastepec (Oaxtepec),
189–190
Meysenbug, Malwida von, 164
Middle East, 8, 10, 15, 22, 25, 188
Milton, John, 91
Missionaries of the Precious Blood, 233
Mithridates vi (ruler of the Kingdom of
Pontus), 140
Monceau, Henri-Louis Duhamel de, 103, 126
Montenegro, 11
Morgan, Frederick, 227
Morocco: 15, **36–37, 38**, 39; Agdal (near
Marrakesh), 36–37; Essaouira, 37; High
Atlas, 37; Marrakesh, 36–37; Ourika
valley, 36; Tangier, 15
Morton, Julius Sterling, 176–177
Muffet, Thomas, 218

Neanderthals, xi
Netherlands: 's-Graveland, **120**
Nietzsche, Friedrich, 164–165
Niger, 15
North Africa, 18, 22, 37, 64, 115
Norway: Sognefjord, 247
Nukarribu (Assyrian gardener), 35

Oberdieck, Johann Georg Conrad, 208
Oman, 15
Ondaatje, Michael, 201
Orient, 59, 63

Palmer, John, 4th Earl of Selborne, 46
Peters, Charles M., 187
Phaeacians (mythological ancient people), 55

Pissarro, Camille, 219, **231**
Platt, John James, 226
Pliny the Elder, 63, 64, 72, 109, 139
Pliny the Younger, 67, **68**
Poland: Silesia (former German province), 130
Pollan, Michael, xii
Pomona (goddess), 59, 65, 115, 163, 215
Pope, Alexander, 55
Portugal: 11, 154, 242; Abrantes, 11; Mouchão, 11
Preece, John, 240
Prestele, Joseph, 176
Prince, William (nurseryman), 171
Puabi (Mesopotamian queen), 44

Qatar, 17
Quetzalcoatl (Aztec god), 189
Quintilian (Roman orator), 86

Rathnayake, Abeyrathne, 200
Redi, Francesco, 159
Rée, Paul, 164
Renenutet (Egyptian goddess), 29
Renoir, Jean, 222

Renoir, Pierre-Auguste, 220, **222**, 223, 224
Rhine (river), 133
Rilke, Rainer Maria, 217
Rome, 83
Romulus and Remus, 73
Rosellini, Ippolito, 42–43
Rosenblum, Mort, 11
Roth, Philip, 184
Rousseau, Jean-Jacques, 150
Russia: 8, 45, 225, 226; Kursk, 45; Topolyevka, 46